# SEAWEED
in Agriculture and Horticulture

# SEAWEED
# in Agriculture
# and Horticulture

---

W. A. STEPHENSON

*With a Preface and updating Appendix by*
ERNEST BOOTH

Second Edition
EP Publishing Limited
1973

This edition offset by EP Publishing Limited
East Ardsley, Wakefield
Yorkshire, England

by permission of the original publishers,
Faber & Faber Ltd.



First published in 1968; this second edition
includes an updating Appendix, *Recent research results*, by Ernest Booth

ISBN 0 85409 848 8

Please address all enquiries to EP Publishing Ltd.
(address as above)

*Printed in Great Britain by*
The Scolar Press Limited, Menston, Yorkshire

W. A. Stephenson

# PREFACE

At some stage in life most of us dream of what we might have done —if only our circumstances had been different. Ambition dies young but we still dream. If we lack the courage and conviction to change, we lack no interest in the few who succeed. Tony Stephenson's story follows simply from his desire to escape from an industrial city and live a useful life in the country. This was the beginning of a fascinating story; the rest is an account of a man with a sublime belief in himself and the new seaweed products which he pioneered.

It seems improbable that an accountant should make such a change at forty and start a new venture on slender resources. The fact that Tony was crippled by a spinal injury at birth should have been an added restriction and the almost complete absence of scientific support for the new uses of seaweed would have discouraged a more prudent man. But, despite his training and disability, Tony was not a prudent man.

Seaweed had been used in animal feedingstuffs since the early days of this century but the idea of making a liquid fertilizer from seaweed was novel in 1950. To make the proposition more difficult, the product was not offered at first to the amateur gardener but to shrewd commercial growers and farmers. In short, both the man and his product were doubtful starters and yet, in the last sixteen years of his life, Tony Stephenson launched his venture and achieved a world-wide distribution for his products. The book records the facts of these years without giving much insight into the man who undertook this apparently rash adventure.

Viewed in retrospect, only a starry-eyed dreamer would have considered marketing a liquid seaweed fertilizer in 1950. Tony was no dreamer, he was essentially an Edwardian character and a most realistic person. His thoughts were direct; he made decisions quickly —and most of them were correct. What better way to gain a knowledge of horticulture than buying a market garden? This was his simple, fundamental approach. He realized his mistake when the books were 'in the red' and he dropped the venture. The experience, however, proved invaluable in selling his fertilizer.

His approach to the choice of a house was equally simple. He would have to travel widely and a good train service was essential so he plotted the main railway lines and settled in the middle of the triangle formed by the old G.W.R., L.M.S. and L.N.E.R. routes. The lack of research into the value of liquid seaweed fertilizers did not worry him; after all, stable manure is good, but the agricultural scientists say little in its favour. In all these matters he was severely practical.

His greatest gift was the ability to 'switch off', as he would say, and turn to his main pleasures—conversation and people. In fact, he never 'switched off' but merely 'switched over' and pursued a new topic with typical zest. Probably his greatest asset was his genuine love of people; he was really interested in people and their troubles and successes. He sought 'characters' and new friends as other men collect stamps and he gave even more readily than he received. Not only was he interested in people, he was an interesting person and this was his strongest point.

These were the qualities that launched the new business; a good product made it expand and invited competition. It was a happy, industrious life; never easy but never dull. Even the end came in the midst of pleasure as he was overwhelmed by a sudden flood whilst swimming in the River Cover in Yorkshire on 4th September 1966.

In 1950, seaweed was a curiosity. Today almost a million tons are collected annually; tomorrow it will be farmed like other agricultural crops—indeed, it is already farmed in China and Japan. Tony Stephenson played a part in the revival of this industry and his book is a piece of economic history; an interesting romantic period piece like its author.

ERNEST BOOTH

ACKNOWLEDGEMENTS

I have been helped in the writing of this book by Mr. Arne Jensen of the Norwegian Institute of Seaweed Research; Mr. R. H. McDowell of Alginate Industries Ltd.; Mrs. Angela Bates of Vitrition Ltd., Castle Street, Stamford, Lincs.; Mr. Ernest Booth of the Institute of Seaweed Research, Inveresk, Midlothian; Mr. Jeffery Waggett, Cwmconnell, St. Dogmaels, Cardigan; Mr. T. Rook, Great Bookham, Leatherhead, Surrey; Captain A. W. Kidner, Lakenheath, Suffolk; Mr. W. D. Howard, Jesus College, Cambridge; Mr. J. Temple, Bluegate Nurseries, Pyrford, Surrey; Mr. J. C. Habgood, Stock, Essex; Mr. L. C. Chilcott, Brent Parks Manager, London; and Mr. A. G. Roberts, Halstead Hill, Cheshunt, Herts.

Acknowledgements are made for the use of tables to: Dr. W. A. P. Black and Dr. F. N. Woodward (1, 2, 27, 28, 29, 30, 32); Alginate Industries Ltd. (6); Professor G. E. Fogg (7, 8); Dr. R. F. Milton (9, 10); American Society for Horticultural Science (12); Clemson College of Agriculture, South Carolina (13, 14, 15, 16, 19, 26); Norwegian Institute of Seaweed Research (3, 17); Connecticut State Highway Department (20); Mr. Martin Austin (22, 23); Bahrein Department of Agriculture (34); Henry Doubleday Research Association (27, 28).

W.A.S.

---

I am extremely grateful and most indebted to my husband's friend and colleague, Mr. A. K. Astbury, for all the help he has given me since my husband's death, both in the final editing of this book for publication and in seeing it through the press. His editorial advice and assistance have been invaluable.

WINIFRED M. STEPHENSON

*Holdenby, Northamptonshire*

# CONTENTS

# ILLUSTRATIONS

## PLATES

## TABLES

# 1

# A PERSONAL NOTE

The reader may welcome a short introductory account of how I became interested in the agricultural use of processed seaweed. Such an account of my own experiences—or rather of our experiences, since they have been shared, in all respects, by my wife—may also help towards an understanding of the subject itself.

I am by profession a chartered accountant, and spent the whole of the war, and the years immediately after, in helping to run an engineering business in Birmingham. Twelve years of industry, however, proved more than enough for me, so my wife and I decided to look for an occupation which would make it possible for us to live in the country.

It was at this time that I became friendly with a consulting biochemist, Dr. Reginald Milton, who had been concerned with the industrial use of seaweed during the war. At that time, in 1944, it seemed possible that the Japanese advance in north-east India might cut off our supplies of jute, which were used in vast quantities for making camouflage netting and other essential products. It was thought that seaweed might prove to be a substitute for jute—at least in netting, where tensile strength is unimportant. It was Dr. Milton's task to carry out research on these lines.

In the event, although factories were set up to process the seaweed in western Scotland, the idea proved impracticable. Not only were there production difficulties, but the seaweed camouflage netting actually dissolved in the rain—or, to be completely accurate, much of the calcium and beryllium alginate fibre intended for use as netting had to be abandoned because sodium alginate, an impurity which still remained in the fibre in spite of stringent precautions, rotted in a wet climate.

But initial research into the possibility of producing textiles from seaweed brought Dr. Milton to a research unit set up in a country house near Birmingham. This house had large gardens and glass-houses; and in his spare time Dr. Milton, who is a keen gardener, tried out various ways of treating seaweed so that it could be used in horticulture. It was during the course of these experiments that he hit on the idea of making seaweed soluble so that it could be given as an easily assimilated plant food. The seaweed solution he produced was the forerunner of the liquid seaweed extract my company now markets under the name of Maxicrop.

We met Dr. Milton after the war, and it was while discussing with him our plans for a new life, away from cities, that he suggested we should manufacture and market seaweed fertilizer in liquid form. It was an attractive proposition for a number of reasons. For one thing it gave us a chance of starting something from scratch, and without any precedents. No one knew anything about liquid seaweed fertilizers; and apart from my wife's botanical training, neither of us knew anything about gardening. I had the basic commercial and accounting knowledge necessary for running a business; but neither my wife nor I knew anything about selling. To start an enterprise technically new, and commercially unorthodox, was for us an exciting challenge.

Further, to carry out our plans it would be possible—indeed necessary—to live in the country away from towns and industry. It would also be necessary, in learning about growing things and the application to them of seaweed fertilizer, to concern ourselves essentially with living things, from soil bacteria and plants to cows. As events have proved, it has also given us an excuse for visits to, and holidays in, such pleasant places as Galway, County Mayo, South Uist, the Orkneys, the Bay of Biscay, Jersey, Guernsey and Wester Ross.

There was one final reason why we were interested in seaweed, and here we were prompted by motives which were partly commercial and partly idealistic. It was already clear that the world would need more food. At the same time, the countless millions of tons of valuable organic material in the sea were largely unexploited, both as a source of fertility to the soil and as a source of nourishment to animals—and, through animals, to men. We felt it must be right, and commercially profitable as well, to exploit this last large-scale natural resource. We felt, too, that the trend towards chemical farming had gone too far, and that in time there would be a reaction in favour of organic manures, of which seaweed was one. Thus we hoped that in developing seaweed we would be in the happy situation of doing something useful and making a profit at the same time.

But if we were to go in for the processed seaweed business we had first to learn something about the products we proposed to sell: and that meant getting some experience of commercial growing by running our own market garden.

As things worked out, however, we produced our first liquid seaweed extract for use as a fertilizer, albeit only on an experimental scale, while still in Birmingham. Dr. Milton had made his first soluble solutions of seaweed by applying steam heat to seaweed and water, held under pressure in a closed vessel. The vessel was heated with a steam jacket, or had a steam coil passing through. When I decided to enter the seaweed business, Dr. Milton made this basic laboratory method over to our first company, Plant Productivity Limited.

It then became necessary to adapt the process to factory production. I had already decided that even if I did go in for processing seaweed, I was not going to have another factory round my neck. In any case I could not run a factory and develop an entirely new project at the same time. I therefore looked around to find a plant where liquid seaweed extract could be made for me.

In the summer of 1949 all was ready for our first attempt, and a batch of 40 gallons was made for us in a north London plastics factory. This we used on our garden in Warwickshire, and gave away to friends. At that time we had not decided on a name for our new product; but Dr. Milton and I had a short list of suggestions, and in the period before we began production in earnest, we sat down to study this list with a bottle of brandy between us on the table, determined not to get up until we had chosen a name. When we finally staggered to our feet, the name was chosen—'Maxicrop'.

Commercial production began in a factory at Slough; the first of the batches was produced in March 1950, after we had left Birmingham.

But I am anticipating. We must go back to the point where we were considering setting ourselves up as market gardeners. Our first step was to sell both our share of the Birmingham engineering business and our house in Warwickshire. This we did in October 1949, and from then until the end of January 1950 we stayed with friends in County Durham. It was during this time—when I was also making arrangements to manufacture in Slough—that I began negotiating for a nursery in Lincolnshire.

Our choice fell on a small commercial nursery at Blankney, between Sleaford and Lincoln. It was some four miles from the edge of the Fens, on a stretch of wonderful deep loam running north and south between the stony, limy heathland to the west, and the Roman

Car Dyke to the east. It had about 13,000 sq. ft. of glass, and 9 acres of open land. I will describe our work at the nursery first from a commercial point of view, if that is the correct term in the circumstances, and then our operations when we used the nursery as trial ground for our seaweed venture.

We grew a wide variety of crops—certainly too many for commercial wisdom, as we later discovered. They included tomatoes, both indoor and outdoor, cucumbers, chrysanthemums, cyclamen, cineraria, *Azalea indica*, forced bulbs for the Christmas trade, early potatoes, cauliflowers, cut flowers, cabbages, vegetable marrows, marigolds, lettuce and broad beans. We also tried, unsuccessfully, to grow peppermint, from which peppermint extract is made, and foxgloves, from which manufacturing chemists make the heart drug digitalis.

The whole thing was quite unprofitable, and most of my time was spent tearing round the countryside trying to sell the stuff we had grown. The moment for decisive action finally came at Christmas 1951. Covent Garden was on the floor. The prices we were getting for our azaleas were barely enough to pay the Dutch supplier of the stock, let alone show us a margin for our work. We had already decided we could lose £2,000 on the venture without being financially crippled, regarding this as payment for experience in commercial growing, and the application of seaweed to it. But the losses which faced us that Christmas were too disastrous to accept. We therefore had our Land Rover and trailer fitted with shelves, and I called on every florist within twenty miles. As it happened, I sold the whole stock from the nursery, and could have sold more. But it was already too late; and the effort, though exhilarating for once, was hardly an exercise to be repeated at regular intervals. The end of our £2,000 was in sight, and we left Blankney on Lady Day, 1952.

Now a word about our experiences with seaweed extract as a fertilizer. The first batch from Slough was delivered to us at Blankney about ten days before we began work at the nursery—in fact on the very night we arrived at the house. It was subsequently bottled by our nursery staff.

We had already engaged an orthodox commercial manager and staff—orthodox in the sense that they knew nothing of the use of seaweed in horticulture. We realized that our first task was to convince them of the value of our product. If we could not convince them, we would certainly not be able to convince the public.

Our first undertaking at the nursery was to prick out thousands of tomato seedlings, using various strengths of liquid seaweed extract to see what they would stand. I found to my surprise—and satisfaction—that the seedlings could be dipped in neat solution and planted out at once without suffering any harm. And so the staff, three weeks after planting, had something to go on. Incredulous at first, if not sceptical, they were beginning to wonder whether there might not be something in seaweed after all. We have continued to use the extract, and with satisfactory results, ever since.

I have said already that commercial production of liquid seaweed extract began at Slough. The factory was one which concentrated on the making of all kinds of equipment and apparatus for the chemical industry, particularly rubber-lined tanks, pipes and valves. The company did no processing of the kind that we required. Nevertheless they agreed to construct, and install, an autoclave, a pressure vessel in which liquid seaweed extract could be prepared according to Dr. Milton's method.

Not surprisingly, the bench process did not work when transferred to the factory. Instead of being presented with 140 gallons of dark, clear liquid, we were given a mixture of liquid and large, sticky, unmanageable lumps of sludge. Not only was the product not a proper liquid extract, but it also frequently blew up with a loud report. This caused pain, embarrassment and alarm to the producers, and called for the deployment of all my powers of mollification. One drum, on its way to our first overseas customer in the United States, blew up in the hold of the ship, causing damage to the surrounding merchandise. Its owners complained. Another bottle blew up in a London store, scattering its contents over nearby foodstuffs. I need hardly say that the liquid extract we now sell is not subject to these disadvantages. We think the earlier batches exploded because the formalin (used as a preservative in this, as in many organic liquids), did not penetrate into the sludgy masses resulting from incomplete hydrolization of the seaweed. As a result, after a few weeks in the can fermentation took place in the centre of the mass. This spread throughout the material unaffected by formalin, and the gas so produced blew up the container.

It soon became clear that the company at Slough, which had no works chemist and did the job for us as a favour, could not be expected to deal with these contingencies. We had to look for another processer. This, to our good fortune, we found in the English Grains Company Ltd., of Burton-on-Trent. This company had all we wanted

—chemists, processing staff, high-pressure steam, ample space, its own transport fleet. We moved the process there in 1953, when our requirements were some 10,000 gallons a year or less. Now, in 1964, we need upwards of 200,000.

We first tried selling the extract through distributors who handled other products as well. As we could offer little practical evidence of the value of what we were selling, knew nothing about writing sales literature, and depended on distributors and salesmen who had other things to push, it is hardly surprising that our first sales campaign did not succeed. It was not until a year had elapsed after the appointment of our first general distributor that we decided we must have our own sales force.

There were still, at that time, a number of men who had been in the forces, and come out with a gratuity and a determination to be their own masters. We were fortunate in finding a number of such men, who knew enough about agriculture and horticulture to sell seaweed products for us. As a result we now have, in addition to many active distributors with their own salesmen, six full-time salesmen of our own, several part-time salesmen who also sell other things, and an export manager who looks after the distribution of our seaweed extract all over the world.

Yet the setting up of our own sales force might never have been possible without another development which occurred while we were at Blankney: this was the opportunity to market seaweed meal, as well as liquid seaweed extract.

Seaweed meal was already being produced in this country during the First World War, and even earlier in the United States, but as far as we were concerned the story began in 1950, when a Mr. Alick McInnes set up a small seaweed factory at Nairn, on the Moray Firth. As a child he had fed seaweed to the cows on his grandfather's croft. When he returned to Scotland after a career as a banker in India, he decided to produce dried seaweed meal as stockfeed and fertilizer. He and I were at that time the only persons active in this field. We soon got wind of each other, and I went to Nairn to meet him in January 1951. As a result of this meeting it was decided that I should market his seaweed meal for him, and I formed for this purpose a company under the title of Seaweed Agricultural Ltd. Thus it came about that we began selling two varieties of dried seaweed meal, one for stockfeed and one for manure, under the name of 'Neptune's Bounty'. At the same time our liquid seaweed extract was being marketed by Plant Productivity Ltd., although we sold this company in 1952 and formed Maxicrop Ltd. in its place.

The marketing arrangement with Mr. McInnes was a useful one, because it enabled us to form our own sales force. We could not, at that time, have employed salesmen full-time on the sale of 'Maxicrop' alone. The whole marketing approach, if we had sold only liquid fertilizer, would have been too seasonal to keep salesmen in all-the-year-round employment. Nor had we then any idea of the possibilities of seaweed extract used as a foliar spray—that is, applied to the leaves of fruit trees, tomato plants and so on, as immediately available nutrient. This technique has since vastly increased the demand for seaweed extract.

But with the addition of seaweed meal to our business, it became possible to employ salesmen throughout the twelve months. They could sell seaweed meal to farmers from August to February, and liquid extract to growers and orchardists from March to July. The seasons overlap at both ends, but this is a reasonably rough-and-ready division.

There was one further reason why it was useful to have seaweed meal to sell: it was, at first, easier to explain the advantages of seaweed meal than those of liquid seaweed extract. We therefore started our new selling campaign by putting a few advertisements in the farming papers, and sending salesmen direct to farmers. Those we approached first were breeders of Channel Island dairy cows, for we thought them the likeliest buyers: we could point out to them that the Jersey and Guernsey breeds developed on pastures which had been manured with seaweed for countless generations. Seaweed must clearly be good for their cattle.

It is to be admitted that we had precious little experimental evidence at that time to put forward in support of our claims; but there were enough breeders and dairy men prepared to listen sympathetically to allow us a fair range of customers. These, in our first year's marketing of seaweed meal, bought 450 tons from us. At the end of that year we had two full-time salesmen who not only sold seaweed meal but also had an active list of customers for liquid seaweed extract.

Our first real stroke of luck came when Barkers & Lee Smith Ltd., a firm of animal feedingstuff manufacturers with mills at Lincoln, Hull and Beverley, became interested in our seaweed meal. At first they included it in their dairy rations on an experimental basis, but after a while they became convinced of its value and decided to include it in nearly all their breeding rations—not only for dairy cows, but also for sows and weaners, ewes and lambs, hens and horses. Later on, the managing director of Barkers & Lee Smith became a member of the board of Seaweed Agricultural Ltd. His company now shares with us the distribution of seaweed meal, and has supplied the greater part of the finance, as well as the long experience of the nutritional needs of stock, both necessary for the development of this side of our business.

Well over a score of manufacturers now include seaweed meal as a mineral supplement and natural foodstuff in their rations; and a venture which began with one small seaweed drying plant on the Moray Firth now markets meal produced by drying plants in Scotland, Ireland and Norway, and has port stores all round the coasts of Britain.

One of our most valued orders for seaweed meal came in 1958—for 5 cwt. of meal for the Queen's geraniums. I delivered the seaweed myself by Land Rover to Buckingham Palace; and we still have, growing in our glasshouse at Holdenby in Northamptonshire, the gardenia given me by the Queen's gardener who accepted delivery.

The seaweed meal we produce has to be dried within a few miles of the coast because the weight of water makes it too costly to transport far. It is gathered, dried, ground and put into bags of our design by small independent producers who make the meal to our specification. Thus while we can control the manufacture of our liquid seaweed extract at Burton, we can exercise only remote control over the makers of seaweed meal.

I have moved away from Blankney, where we were running our market garden and launching our seaweed business simultaneously. Once we had decided that the time had come to move, we drew a fifty-mile-diameter circle on the map with its centre at Meriden, the village said to mark the middle of England. Somewhere within that circle, centrally placed for our task of building up seaweed production and distribution throughout England, we must make our new home.

The first place which came up was the one-time rectory at the tiny Northamptonshire village of Holdenby, just on the edge of our circle. Holdenby was where Charles the First was imprisoned after his defeat in the first phase of the Civil War. His residence ended when Cornet Joyce rode up to Holdenby House with a troop of horsemen to take him away. 'Where is your warrant?' asked the King. 'Here!' said Joyce, pointing to his men. 'It is as fair a commission, and as well written, as I have seen in my life,' the King replied. And off they went together.

The old rectory, where we are now living, is big enough to house us and our office staff, and provide a flat for my secretary. Its 2½-acre garden is also big enough for the preliminary trials and experiments we wish to carry out. Trials at Blankney had taught us that it was impossible with amateur resources to carry out properly controlled plot experiments in open ground. In the light of this experience we decided that we would do two things only: first, grow as many ordinary vegetables and flowers as possible, so that we could speak with knowledge of their reactions to seaweed; and second, carry out pot-scale experiments which could be controlled without a trained staff, and from which valuable, if limited, conclusions might be drawn. For both these purposes our garden and glasshouses at Holdenby are suitable.

I have purposely made clear my commercial interest in seaweed products. It would be impossible to conceal the circumstance, even if I wanted to. But in the chapters which follow, I shall try as far as possible to avoid making too repetitive a use of our brand names of 'Maxicrop' for our hydrolized extract, and 'Neptune's Bounty' for our meal. Since, however, I must speak of what I know, I should make clear now that unless otherwise stated, all claims and recommendations refer to our own products. We have no reason whatever to doubt the efficacy of other brands of seaweed meal and extract. Where these are used, for example in overseas trials, we shall say so. In all other cases, and particularly in mentions of reported interviews and recommended treatments, it is our own products of which we speak, and for which we take responsibility—and, we hope, credit.

# 2

# WHAT SEAWEED IS, AND WHERE IT GROWS

Wherever sunlight, warmth and nourishment are found in nature, there life will be found also. This applies to the sea as well as to the land; for several million years, seaweed has been the most economical way of exploiting the sunlight, warmth and nourishment found at the edges of the ocean.

The number of different kinds of seaweed found throughout the world hardly concerns us. It can be said, however, that there are forty-two different families of seaweed in British waters, although only three of these families, the Fucaceae, Laminariaceae and Gigartinales, are of commercial significance, simply because they do provide the vast bulk of the material available there. The three individual seaweed species most commonly gathered for industrial and agricultural use are all Fucaceae—*Fucus vesiculosus*, *Fucus serratus*, and *Ascophyllum nodosum.*

Some of the Fucaceae, which grow between high and low tide marks, contain air sacs to help them float up to the sunlight when the tide rises, and are generally known as wracks. The 'true' bladderwrack is *Fucus vesiculosus*—true, that is, in a restricted etymological sense, since although some other Fucaceae have gas-filled vesicles, *Fucus vesiculosus* is the only one which includes the word in its title. *Ascophyllum nodosum*, which also has gas-filled vesicles, is sometimes called 'knotted wrack'. *Pelvetia canaliculata*, another type of *Fucus* with roughly triangular sacs filled with water and soft substance, is called 'channel wrack' or 'calf weed'.

Fucaceae, which form conspicuous masses of vegetation on any rocky western shore of the British Isles, are also to be seen on the coasts of Greenland and Iceland, as well as in the Sargasso Sea, that area in the still centre of the Atlantic where floating seaweed abounds. They are, by contrast, also known to over-winter on rocks in north-west Greenland in latitude 76° 30″.

Most Fucaceae live for a few years, but are usually torn away from their anchorage; not because they have come to the end of their natural life, but because their bulk has become too great to withstand the pull of moving water. *Ascophyllum* has been known to live anchored in Trondheim Fiord in Norway for nineteen years, although it will have lost many of its fronds in that time. One worker in Wales found no *Ascophyllum* over fifteen years old, while on the semi-exposed boulder shore of the inner part of Galway Bay in Ireland, *Ascophyllum* lives no longer than thirteen years.

Members of the Laminariaceae, the second of the three families I mentioned earlier, have no air sacs. They have a longer, thicker stem than most seaweeds, and flatter, wider, single fronds, something like a wide sword blade. Their edges are slightly corrugated.

All seaweeds have certain characteristics in common. They are simple in structure, have no true leaves, stems or roots, and contain no wood or vessels. The whole of their body, which is made up of simple cells with little or no differentiation, is of the same construction throughout.

Seaweeds are not produced from seed. In this they differ from flowering plants, which produce a seed containing an embryonic plant. Seaweeds, in common with other non-flowering plants, reproduce themselves by what is called a spore. Spores have no embryo: the young plant begins to take shape only after the spore has begun to germinate.

Little more about the botany of seaweeds need be said, except to point out that they form part of the botanical order of Algae and are divided into three main types according to obvious differences in colour: *Chlorophyta* (green seaweeds), *Phaeophyta* (brown), and *Rhodophyta* (red). Each order contains families whose names end in -aceae (such as Fucaceae and Laminariaceae, already observed). Each family is then split up, finally, into species which have a second name describing them specifically, such as *Fucus vesiculosus*, a member of the Fucaceae family having vesicles.

Since there is so little variation in the chemical and physical composition of sea water, it is impossible to differentiate seaweeds according to the water in which they grow, at least in the same way in which it is possible to classify land plants according to the differing soils in which they flourish. An exception can be made here, however, where the temperature of the water and the availability and strength of the sunlight are concerned. Both can encourage, or discourage, the growth of particular types of seaweed, so that there are certain weeds which flourish in tropical, or temperate, waters; and some which do not. In general, brown seaweeds grow better in temperate waters, and red seaweed in warmer waters.

The depth at which seaweed grows, whether in temperate or tropical waters, also depends on the amount of sunlight available. Yet even in ideal conditions, where sunlight is strong and vertical—as at the equator—few seaweeds grow in greater depths of water than twenty-five fathoms, except for large weeds whose ends float near the surface. But seaweeds can grow even in the absence of light. They are found, for instance, on rocks under the Antarctic ice, where only 1 per cent of the light which falls on the surface of the ice reaches the plant.

The degree of salinity in sea water has a direct influence on the growth of seaweed. Salinity is constant enough in open seas, although the level falls in and near estuaries, and in such partly enclosed seas as the Baltic and Mediterranean. Growth is generally reduced as salinity decreases—in the Baltic, for example, the same types of seaweed become noticeably smaller as one goes farther north, and are smaller than the same seaweeds growing in the same latitudes in unenclosed seas. Some seaweeds do grow better in less salty water (possibly because of additional nutrients in fresh water); and most seaweeds fail to make much growth where salinity fluctuates considerably, say in tidal pools subject to swift evaporation under hot sun or wind. The effect of salinity on seaweed supplies has, however, little commercial significance.

The chemical characteristics of seaweed, which are important for our present purposes, result partly from the conditions in which seaweed live. It is therefore necessary to describe these in more detail.

The seaweed's 'soil' is the sea, and the physical and chemical characteristics of the sea decide its form and content. Seaweed floats in sea water and does not have to support itself. For this reason it has no need for the fibres, bark or lignin which, by giving a plant rigidity, enable it to support itself on land. Seaweeds give way to the sea; and the violent effects on them of wave action are to some extent countered by the tough elasticity of their cell walls.

But the sea is not only the seaweed's physical support: it is its chemical support also, in the sense that it nourishes it as soil nourishes the land plant. It is true that seaweed, like all land plants, uses sunlight to take carbon from the air (where it appears as carbon dioxide), and then uses this carbon to build up its bodily tissues. But while land plants take their carbon from the atmosphere, sea plants take it from the air contained in the sea itself—the same air, incidentally, which fish breathe. But apart from this process of using carbon and sunlight to create tissue—the process is known as photosynthesis, and is dealt with more fully later—seaweed takes all its nourishment from the sea, by absorbing it through the whole of its surface. The seaweed plant thus combines in its surface both roots and leaves. It is through these surfaces that it absorbs, from the sea, those nutrients which the land plant absorbs, through its roots, from the soil.

For this reason, seaweed has no roots of the same nature as those of the land plant. What might be taken for a root is in fact only a holdfast which anchors the weed to rock, shell, pile, girder, or even another plant.

Not all seaweeds have such holdfasts. The headwaters of many lochs in Scotland are at times filled with freely floating weed, while the bulk of the seaweeds which make up the Sargasso Sea are also free-floating.

Growing conditions in the sea differ from those on land in a number of ways. The sea is more consistent. It is not subject to the wide and often rapid changes of temperature which affect soil and climate over wide areas of the earth, as well as the plants growing there. Nor does the sea ever become arid, as some soils do. Although the concentration of nutrient salts in sea water varies, being low in early summer and high in winter, there is never any real shortage of food in the sea for seaweeds, as there may be in the soil for land plants. Indeed, it is true enough for our purposes to say that sea water contains all known chemical elements, however small their proportions may be.

Sea water is constantly in motion, supplying food, and oxygen, to plants below the surface, and dispersing their waste products. Nor can the sea, which brings food to the seaweed, be impoverished by successive crops of weed, as the soil is by successive agricultural crops. The nutrients in the sea are being replenished constantly by nature; and the nutrients leached out of the fields by rain, which are then lost to agriculture, sooner or later find their way to the sea where, in common with all the other salts and minerals there, they are universally and uniformly available.

To this rule the Dead Sea and other enclosed seas are exceptions which do not affect our argument. Nor do the slight variations which occur in the case of partly landlocked seas such as the Mediterranean deny to the seaweeds which grow there that ample and uniform nutrition available on the Atlantic and North Sea coasts of the British Isles. There are no deficient seas. Seaweeds never suffer from shortage of nutrients; and, living in a stable nutrient medium, are always of stable chemical composition. With one or two known exceptions, none of commercial importance, there is much greater similarity between the seaweeds of the world than there is between the land plants; and the chemical composition of any one type is always more constant.

But although seaweeds live in such close harmony with the sea, the chemical consistency of seaweed does not reproduce exactly the chemical consistency of sea water. For one thing, seaweeds dilute the common salt content of the sea. They also accumulate some of the elements found there to several thousand times their concentration in sea water. Thus the iodine content of seaweed is 20,000 times greater than the iodine content of sea water, which makes seaweed far and away the richest source of iodine in the world. Even Chilean nitrate, which has replaced it commercially as the raw material for iodine extraction, comes from prehistoric seaweed deposits.

Now while it is true that the overall chemical consistency of seaweeds is greater than that of land plants, lesser variations must be noted in their chemical composition. These variations occur according to species, stage of growth, time of cutting and locality. There are even a few freakish seaweeds like *Pelvetia canaliculata*, which has a much higher fat content than other weeds—one amounting to as much as 10 per cent on occasions.

*Corallina*, too, which attracts a coating of lime, has in the past been mistaken for coral because of its hard and brittle construction. But such eccentrics as these are out of character with seaweeds as a whole. Variations which are found in ordinary seaweeds are usually seasonal. Some seaweeds are high in mineral matter and crude protein in the spring, but low in carbohydrates, while in autumn the reverse is the case.

Here are two analyses of seaweed, including the brown (*Phaeophyta*), red (*Rhodophyta*), and green (*Chlorophyta*) varieties.

TABLE 1

Simple analysis of five seaweeds

Analyses are of dry matter content after the material has been dried at between 60 and 80 degrees Centigrade. Figures are expressed as percentages, and an analysis of dried grass is included for comparison. Frond is the 'leaves' of the seaweed, stipes the stem. Nitrogen-free extracts are non-protein material such as sugars.

| | Ash | Protein | Fat | Fibre | Nitrogen-free extracts |
|---|---|---|---|---|---|
| *Cladophora rupestris* (January) | 29·3 | 30·5 | 0·48 | 16·6 | 23·1 |
| *Rhodymenia palmata* (January) | 27·4 | 23·4 | 0·26 | 2·1 | 46·8 |
| *Laminaria cloustoni* frond (November) | 21·8 | 12·6 | 0·38 | 5·0 | 60·2 |
| *Laminaria cloustoni* stipes (January/December) | 37·6 | 8·4 | 0·25 | 9·6 | 44·1 |
| *Ascophyllum nodosum* (December) | 24·5 | 7·6 | 2·62 | 8·4 | 57·8 |
| Grasses | 9–10 | 11–27 | 2·3–3·8 | 22–44 | 38–53 |

This second table shows the mineral content of the same seaweeds. The minerals listed, which appear undifferentiated in the previous table, form only a small number of those in seaweed, although they appear in greater quantity than others not listed.

TABLE 2

Mineral analysis of five seaweeds

The symbols used are: Na, sodium; K, potassium; Ca, calcium; Mg, magnesium; I, iodine; and P, phosphorus. $SiO_2$ is silicon oxide, or silica; Cl′, the chemical symbol for chlorine with a dash added, means chlorine ions. It shows that chlorine is present, but gives an added indication of the closeness of its combination with other elements. The same remarks apply to $SO_4''$, the chemist's symbol for sulphate ions.

| | Na | K | Ca | Mg | I | P | $SiO_2$ | Cl′ | $SO_4''$ |
|---|---|---|---|---|---|---|---|---|---|
| *Cladophora rupestris* | 2·50 | 3·28 | 1·52 | 0·73 | 0·11 | 0·27 | 7·1 | 6·34 | 4·62 |
| *Rhodymenia palmata* | 2·07 | 7·91 | 0·72 | 0·39 | 0·03 | 0·56 | 2·2 | 9·70 | 1·06 |
| *Laminaria cloustoni* frond | 2·88 | 5·25 | 1·04 | 0·58 | 0·50 | 0·28 | 0·6 | 5·92 | 3·06 |
| *Laminaria cloustoni* stipes | 1·35 | 8·15 | 1·80 | 0·73 | 0·33 | 0·25 | 0·5 | 12·48 | 2·54 |
| *Ascophyllum nodosum* | 2·90 | 2·26 | 2·16 | 0·82 | 0·05 | 0·09 | 5·0 | 1·89 | 6·95 |
| Grasses | 0·13 | 1·99 | 0·37 | 0·24 | 0·00005 | 0·21 | — | 0·50 | — |

This third table gives a detailed analysis of *Ascophyllum nodosum*, for our present purposes the most important of all seaweeds, since it forms the raw material both of seaweed meal and of liquid seaweed extract.

TABLE 3

Analysis of *Ascophyllum nodosum* by Norwegian Institute of Seaweed Research

| COMPONENTS | % | CARBOHYDRATES | % |
|---|---|---|---|
| Proteins | 5·7 | Mannitol | 4·2 |
| Fat | 2·6 | Alginic acid | 26·7 |
| Fibre | 7·0 | Methylpentosans | 7·0 |
| Nitrogen-free extracts | 58·6 | Laminarin | 9·3 |
| Moisture | 10·7 | Undefined sugars | 14·4 |
| Ash | 15·4 | | |
| | 100·0 | | |

(Nitrogen-free extracts are made up of non-protein material such as sugars)

(Methylpentosans are complex polysaccharides)

ELEMENTS

| | % | | % |
|---|---|---|---|
| Silver | ·000004 | Nitrogen | ·062400 |
| Aluminium | ·193000 | Sodium | 4·180000 |
| Gold | ·000006 | Nickel | ·003500 |
| Boron | ·019400 | Oxygen | Undeclared |
| Barium | ·001276 | Osmium | Trace |
| Carbon | Undeclared | Phosphorus | ·211000 |
| Calcium | 1·904000 | Lead | ·000014 |
| Chlorine | 3·680000 | Rubidium | ·000005 |
| Cobalt | ·001227 | Sulphur | 1·564200 |
| Copper | ·000635 | Antimony | ·000142 |
| Fluorine | ·032650 | Silicon | ·164200 |
| Iron | ·089560 | Tin | ·000006 |
| Germanium | ·000005 | Strontium | ·074876 |
| Hydrogen | Undeclared | Tellurium | Trace |
| Mercury | ·000190 | Titanium | ·000012 |
| Iodine | ·062400 | Thallium | ·000293 |
| Potassium | 1·280000 | Vanadium | ·000531 |
| Lanthanum | ·000019 | Tungsten | ·000033 |
| Lithium | ·000007 | Zinc | ·003516 |
| Magnesium | ·213000 | Zirconium | ·000001 |
| Manganese | ·123500 | Selenium | ·000043 |
| Molybdenum | ·001592 | Uranium | ·000004 |

OTHER ELEMENTS PRESENT

| | | |
|---|---|---|
| Bismuth | Gallium | Thorium |
| Beryllium | Indium | Radium |
| Niobium | Iridium | Bromine |
| Cadmium | Palladium | Cerium |
| Chromium | Platinum | Rhodium |
| Cesium | | |

Seaweeds, like land plants, contain carbohydrates, but those in seaweed differ in several respects from those in land plants. The most important difference is that carbohydrates in seaweed decompose more readily in the soil, and promote the proliferation of soil bacteria. This is partly because they contain about 10 per cent of simple sugars which are available to bacteria without further decomposition, partly because they contain very little cellulose, which is the main constituent of land plants, and notoriously resistant to bacterial attack. The fact that seaweed decomposes readily also means that it makes fewer demands on soil nitrogen than would decomposing cellulose.

The main carbohydrates in brown seaweed (and these seaweeds are the only ones with commercial significance in agriculture) are mannitol, laminarin, fucoidin and alginic acid.

Sweet-tasting mannitol takes the place of the sugars found in land plants. It represents from 5 to 25 per cent of the dry matter of brown seaweed, depending on the time of harvesting, and the species of seaweed. Laminarin, which can compose as much as 25 per cent of the dry matter of *Laminaria* seaweed in autumn, is the 'starch' of the brown seaweeds.

Of the uses of alginic acid, both in industry and in agriculture or horticulture, we shall have a good deal to say later in the book.

I have given here only a partial description of seaweed contents. Seaweed contains many more things whose significance in agriculture and horticulture cannot be appreciated unless the fundamental principles of plant growth are also understood. For that reason the chapter which deals with auxins, vitamins, trace elements and other complex seaweed contents appears later in the book, after a chapter on plant growth.

Now a word about seaweed resources. We know that four-fifths of the surface of the world is covered by sea, and that the bulk of the vegetation on its shallow margins is greater than that of all the land plants of the world. We know for certain that there are heavy seaweed deposits off the coasts of Ireland, Scotland, Norway, Portugal, New Zealand, Australia, South Africa, Korea, Japan, Iceland, and North and South America. (We know this, for one thing, because people living in these areas have written to tell us so.) But deciding how much can actually be harvested and made commercially available in these areas is another matter. It is understandable that the question remains unanswered, for we do not even know, with any certainty, just how much seaweed is available commercially on our own doorstep.

## *What seaweed is, and where it grows*

The most reliable assessments of seaweed resources have been made in Scotland, Norway and Japan. In Scotland, the Institute of Seaweed Research, a government-supported body which carries out, and sponsors, fundamental and applied research into all aspects of seaweed production, harvesting and processing, estimates that one million tons of seaweed could be harvested in Scotland each year without depleting the supply. This estimate was based on aerial photography in wartime, and post-war surveys based on physical sampling.

Scottish seaweed is not used in agriculture on any large scale. Its main use is in the production of alginates, a commercial and industrial commodity. The collection and processing of seaweed for this purpose are highly organized in the Hebrides, Orkneys, Shetlands, and on the north coast of Caithness.

Even so, in spite of large unexploited resources, Scotland now imports seaweed, mainly because of local transport and labour difficulties. Regular transport from seaweed-bearing coasts is either too costly or non-existent; and in two areas, at least, there is not enough labour to collect and dry the seaweed.

The Norwegian Institute of Seaweed Research estimates that Norway has a total standing crop of over a million tons of *Ascophyllum nodosum* alone, although not all of this could be harvested economically. This estimate was based on charts and sampling. Charts of the area surveyed were covered with parallel lines 1,000 metres apart, running north and south. It was assumed that each line covered a band of land, shore and sea 2 metres wide, so one five-hundredth of the whole area of islands, skerries and coastline was thus covered by lines.

Areas between high and low tide marks crossed by the lines were called transects, and were used as sampling areas. Within each of these areas the extent of the intertidal zone, and the degree to which it was covered with *Ascophyllum nodosum*, were measured. Account was also taken of the presence, or absence, of the other main types of seaweed growing in the zone. At every tenth transect all the main types of seaweed, including *Ascophyllum nodosum*, were cut and weighed.

With these measurements, and sampling, it became possible to assess the total length of tide lines in the area, the length of the seaweed zones, the area of *Ascophyllum*, and the total quantity, in tons, of the most important species. The Institute believes, that fully exploited, this *Ascophyllum* could produce 100,000 tons of dried seaweed meal a year.

## *What seaweed is, and where it grows*

Seaweed production in Japan is far and away larger than the total production of all other countries in the world. The F.A.O. *Yearbook of Fishery Statistics for 1964*, which gives world tonnages of seaweed landed between 1961 and 1964 as 608,000, 704,000, 600,000 and 552,000 metric tons respectively, gives Japanese production in 1964 as 360,300. Individual countries' totals are set out in the following table:

TABLE 4

World seaweed production in tons

| | 1961 | 1962 | 1963 | 1964 |
|---|---|---|---|---|
| Morocco | 3,400 | 1,900 | 2,100 | 2,300 |
| Sudan | 700 | 900 | — | — |
| Tanzania | 300 | 300 | 100 | — |
| Canada | 18,500 | 23,500 | 22,500 | 19,800 |
| Mexico | 15,600 | 21,300 | 19,400 | 23,300 |
| United States | 3,200 | 2,500 | 3,300 | 2,000 |
| Argentina | 2,000 | 1,800 | 1,700 | 2,000 |
| Formosa | 900 | 1,000 | 800 | 800 |
| Japan | 424,800 | 501,700 | 425,700 | 360,300 |
| Korea | 40,000 | 51,700 | 42,900 | 42,800 |
| Philippines | 100 | 100 | 100 | 100 |
| Ryukyu Islands | 100 | 100 | 100 | 100 |
| France | 171,900 | 18,300 | 75,400 | 12,900 |
| Norway | 72,100 | 71,000 | 56,900 | 63,600 |
| Portugal | — | 3,100 | 4,200 | 4,800 |
| Spain | 4,400 | 4,900 | — | — |
| Britain | 18,600 | 18,400 | 14,500 | 18,400 |

This list is, of course, an interesting mixture of fact and fiction. The figures for France are not included in the total because their nature is not clearly defined. Some countries which have a seaweed industry are omitted, for instance Chile, whose total seaweed harvest in 1963 was 3,500 metric tons, of which 1,868 were exported. It is also curious that the neck-and-neck production race between the Philippines and the Ryukyu Islands to the south of Japan shows no sign of being resolved over four years. It is difficult to think that these figures, which give each a production of 100 tons annually, neither more nor less, can be accurate; or that the considerable annual variations in French production reflect the total amounts of seaweed harvested.

It is also reasonable to suspect that other countries which possess seaweed resources are missing, or included, for arbitrary bureaucratic reasons—missing perhaps for reasons of security, present for reasons of prestige. Finally, the fact that China is not a member of the United Nations explains why this country, with its vast seaboard and doubtless vast seaweed resources, is credited in the original list with a production of 800 tons. This was, of course, the production of the island of Formosa, as here listed. But how near to, or how far from, the correct world total of seaweed landed, these figures do at least show Japan's pre-eminent position as a seaweed producer.

This pre-eminence results partly from the fact that seaweeds are cultivated in Japan on a wide scale. Not only are bamboo and brushwood planted in shallow seas to catch spores, but nets with 15-cm.-square meshes made of coconut palm, hemp palm and plastic are used for the same purpose, as well as 'blinds' made of split bamboo connected with ropes at 10–15-cm. intervals. Seaweed beds are also created, or improved, by concreting over rocks, by throwing boulders or concrete blocks on to the sea bed, or by blowing up already existing stretches of rock. Cultivating and harvesting implements are also used on a much wider scale than elsewhere.

Seaweeds are grown in Japan for food, for the production of agar and glue, and for other industrial uses. Average annual production of uncultivated seaweed between 1955 and 1960 totalled 270,200 tons; of cultivated, something short of 87,000 tons. It was also reported in 1961 that nearly 68,700 fishermen were engaged in producing cultivated seaweed.

Although Japan is the only country where there has been any large-scale attempt to exploit the seaweed resources of the ocean, evidence of increasing interest in seaweed comes from Argentina and Brazil, where new government laboratories have been set up for its study. A government laboratory in Portugal, as well as two marine laboratories in India, have extended their investigations to seaweed utilization. A department to study seaweed has been set up at the Universidad del Norte in Chile, while the Marine Resources Institute of Santiago is also exploring the economic possibilities of Chilean seaweed resources. There are factories in Spain, Morocco, Norway, Tasmania, the United States, South Africa, Nova Scotia, France, Japan and Scotland for the production of alginates from seaweed.

This is a brief analysis of such information as is available on the world's total seaweed resources, on the amount industrially processed, and on places where research into seaweed is taking place. We have purposely made no reference to the use of unprocessed seaweed in farming, which occurs on a large scale in Ireland, the Channel

Islands and France, and elsewhere. Statistics here would be of little value, and are in any case not available.

We will now consider the harvesting and processing of seaweed in the British Isles; and here and elsewhere crave the indulgence of our friends in the Irish Republic for including their country in this description. I am not myself attached to the designation 'British', which serves the English no better than the Irish; but in this connection the term is useful, and it would lead all of us into difficulties if we tried to avoid using it.

Here, and indeed throughout the world, a distinction must be made between the total amount of seaweed growing in an area, and the amount which can be harvested commercially. Commercial harvesting depends on ease of access, on tides, transport, labour, and other factors, factors which I had in mind when earlier in this chapter I referred to three types of seaweed of commercial value in the British Isles. In singling these out for special mention, I did not mean to imply that others were of no value. There are, I believe, no seaweeds which are valueless to man. What I did mean was that other seaweeds were available in smaller quantities, or for shorter periods; or owing to size or inaccessibility were too difficult to harvest economically; or because of their high water content, could not conveniently be handled.

Subject to these limitations, however, it is the suitability of the coast itself for growing all, or any, seaweed—rather than the presence there of one prolific variety—which has first influence on whether the crop is worth harvesting. The strength of the waves, and the distance between tide marks (which are partly a matter of vertical tidal range, and partly a matter of the slope of the coast) are the two most important factors here. Given a constant vertical tidal range, the width of the weed belt on the shore is then decided by its gradient. It is widest where the gradient is gradual, narrowest where it is abrupt.

The strength of the waves is another, purely local, factor which affects distribution. If the pull of the waves is excessive, few seaweeds (apart from those with tough, leathery fronds) manage to grow, and they tend to be smaller than usual. Even in less exposed places, constant movement of the water may restrict their growth. Since sand and shingle also make the growth of seaweed impossible, the ideal situation is therefore an arm of the sea, sheltered from direct wave action, and having clear rocks sloping gently from high to low water mark.

One effect of tidal range is worth mentioning here as a matter of interest, although it has slight influence on the value of a crop: the tendency for the tide to 'sort out' the various types of seaweed growing on the shore. This it does by varying the growing conditions for seaweed from the top of the tide, where submersion may be momentary and exposure almost complete, to the bottom of the tide, where submersion is complete and constant. Since for each type of seaweed there is, in theory, an ideal period of exposure and an ideal period of submersion, and since these periods vary from seaweed to seaweed, and the change from complete submersion to complete absence of submersion is gradual, this means that tidal movement does in fact sort out the seaweed growing on any shore. At the top of the tide there will be seaweeds which require most exposure to the air, at the bottom those which want least. Thus *Fucus spiralis* grows in a band towards the top of the tide, *Fucus serratus* in a band at the bottom; and other wracks form more or less parallel bands between.

Of more importance from a commercial point of view is the fact that if tidal range is restricted, as it is in the Baltic and Mediterranean, it will also cut down the amount of seaweed which can be easily harvested: there is then less room in which the seaweed can grow. That is why the harvesting of seaweed in these two areas has been, and must be, on a smaller scale than on the Atlantic coasts of west and north-west Europe.

Other conditions are responsible for differences in the amount of seaweed which grows on the northern and western coasts of the British Isles and on the southern and eastern coasts. There is more harvestable seaweed to west and north because the coasts there are rocky. The lower temperature of the air, by comparison with that of our southern coasts, also allows seaweed to grow higher up the beach—on south-coast beaches the main belt of seaweed lies almost wholly below half-tide level, largely because the air, with its greater drying power, makes for over-aridity above this point. Farther north, however, seaweed may reach to the level of the highest spring tides in an almost continuous belt.

A final word about the purely commercial aspects of harvesting seaweed. This is, in general, a part-time business. The seaweed used for processing in the British Isles is gathered off the rocky coasts of west Ireland and the coasts of North and South Uist and Lewis in the Hebrides by crofters who work in their own time to suit the tides, and are paid by the drying factories on a tonnage basis. The scale on which they operate varies. Small farmers on the west coast of Ireland bring—or brought—their seaweed up from the beach on the back of a donkey, and pile it by the side of the road for collection by the factory lorry. Crofters in a larger way of business in Uist use their own lorries, or tractors and trailers, to bring in the seaweed they have collected.

In some remote areas where farm land is poor, and divided into small units, the income from seaweed gathering may make all the difference between the death or survival of a local community. In other regions it forms a valuable addition to family income. For those whose time is not wholly absorbed by farming or fishing, and for those who are more at home with a scythe than with elaborate power machinery, it is still possible for a father and son working together to earn £60 a month by part-time seaweed gathering. If they work really hard they may even double this figure—although the work is of course seasonal, being mainly restricted to the spring and summer months.

Seaweed gathering is therefore an independent business. It is done by independent contractors, if they may be so called, who choose their own time to work, and even choose whether they will work or not. Profits may not be high, but cheap labour is certainly not involved—least of all in Norway, where seaweed harvesting is done by fishermen with a high standard of living.

In conclusion, I must draw a distinction between weed grown for the sake of being harvested, processed and made commercially available, and the vast amounts of cast weed found on the south coast of England and on other coasts throughout the world. Only the former is included in my survey of seaweed resources. Cast weed, torn by tide and storm off coastal rocks and even rocks out at sea, and then washed in heaps on the shore, can be of value to local farmers. Composted, it would prove useful to gardeners who went to the trouble of carrying it away. But it cannot be 'harvested' except in a limited sense, and is hardly capable of general exploitation at all.

For one thing, supplies are unreliable. There is no controlling the tides which wash the seaweed on the shore. Nor is there any controlling the storms which wrench the seaweed off its anchorages up to fifty miles away for final delivery on a tide-washed shore. Some weeds grow far out at sea on reefs which, while always submerged, are near enough to the surface at low tide to support plant life. I believe that some of the seaweed washed up on the south coast of England may come from shallow areas in mid-Channel, perhaps from the reef halfway between Dungeness and Cap Gris Nez known to the French as 'Le Colbart' and to us as 'The Ridge', where the water can be as little as 6 ft. deep at low water.

The vast heaps of weed found on the south coast of England may have been torn from their anchorages some months before. Some of the weed may be fresh; most will have been dead for some time, and tide and rain will have washed out some of the nutrients it contained. Lack of freshness may be of little moment where weed is carted straight from the shore for spreading on the land; but such weed, of infinitely variable composition and quality, is not suitable for drying and grinding into seaweed meal for use as human or animal feeding-stuff, or as fertilizer. Even less useful is it as the raw material from which liquid extract can be made. For these purposes seaweed should be freshly cut, and processed within a limited period. Further, cast weed may often be badly damaged, broken into small fragments, adulterated by oil and other jetsam, and mixed with sand. This might matter little where weed was spread locally without treatment; it would make impossible the use of such weed for wider commercial or agricultural use.

So the fact that cast weed is never fresh, clean, or of uniform quality, automatically rules out its use for commercial processing. And, as I have already pointed out, even if the quality were acceptable, the supply is unpredictable—negligible at calm periods, vast after storm. For commercial processing, a regular and reliable supply is essential to keep transport and plant occupied. It is useless to hope for such a supply from cast seaweed.

As far as the south coast of England is concerned, access of vehicles to beaches is also restricted, and mechanized collection difficult because of steeply shelving shingle and other obstacles. In areas where large supplies of cast weed are at times available, the labour costs of collecting and delivering seaweed to a local plant, and of processing it there, might be prohibitive. And doubtless these factors operate on other coasts as well.

It is still a fact, though, that over large areas of the world freshly cast weed is still of immediate local value, and what was true in 1840, when Asenath Nicholson made a passing reference to picking up cast weed from the beach in *The Bible in Ireland* is still true today.

'Next morning,' she writes, 'the tempest was still high, and venturing upon the strand, I there saw, as at Valentia, crowds of females busied; and speaking to one, she replied, "These stawrmy nights, ma'am, blow good luck to the poor; they wash up the say-weed, and that's why ye see so many now at work." The company increased, till I counted more than sixty; and busy, merry work they made of it; running with heavy loads upon their heads, dripping with wet, exultingly throwing them down, and bounding away in glee.'

# 3

# THE TRADITIONAL USES OF SEAWEED

Seaweed is the last large-scale natural resource unexploited by man. There are a number of reasons why this should be so. The first is that men have always assumed that seaweed, if it had any useful function at all, had at best only a subsidiary one—as an additional form of organic manure for use in farming. And until recently, alternate sources of organic material were already available in the form of natural animal manures, mainly from the farmyard. These were still easily obtainable, and on a large scale, until the late 1920's. Even after this time, when it became more difficult to find good organic manure, organic residues remaining in the soil after years of treatment with farmyard manure were still enough for farming needs.

By the end of the Second World War, however, these residues were becoming worked out—and there was not enough farmyard manure to replace them. With a new and unsatisfied demand for organic manure, the main excuse for our failure to exploit seaweed on a large scale was no longer valid. Other reasons why seaweed was neglected before, except on a local scale, were not as compelling, but are easily guessed at. Seaweed is not easy stuff to harvest, and it usually grows where labour is scarce. It is not always readily accessible, and its harvesting must be planned with reference to the movement of the tides. Nor is it easy to handle, being wet, slippery and bulky. Further, to reduce it to a form in which it can be stored and distributed commercially, it must be dried mechanically, and perhaps further processed. Moreover, it is only in recent years that drying and processing techniques capable of treating seaweed have been developed —and even where development was feasible, have been applied. For it did not seem to occur to anyone to think seriously of exploiting seaweed; it had been used for centuries, all over the world, in unprocessed form by farmers living near the sea—and the next step, its processing for the benefit of growers everywhere, was not taken until very recent times.

The history of the employment of seaweed in farming and industry, to which I have referred in passing, is a well tilled one. There is little point in going over it again, save to say that the Romans knew of the value of seaweed as an animal feedingstuff, and as a manure for vegetables. They also used seaweed for dyeing; Pliny refers in his *Natural History* to a seaweed dye used in Crete which was so 'fast' that no amount of washing affected woollens coloured by it. Brown seaweeds were also once used to produce a type of red dye in Wales (although modern experimenters have been unable to confirm this use).

In more recent times there are many references in, and after, the sixteenth century to the use of seaweed manure in the British Isles, particularly on sandy soils in Scotland. Barley, potatoes and oats, as well as many vegetable crops, were all said to profit from it.

Fresh seaweed is still used as manure. As far as the British Isles are concerned, it is utilized on farms near the sea for two distinct purposes; where there is hardly any soil at all it comes to the rescue, and where the soil is already rich and high yielding it makes it better still. In the first case, where absence of soil may or may not be associated with subsistence farming, it is used to create soil where none existed; in the second, to enrich soil which is already in very good heart.

It is in areas such as the Aran Islands, off the west coast of Ireland, that seaweed is used to 'make' soil. The seaweed is mixed with sand or sandy soil, and the resulting mixture, often laid on top of bare rock, makes what are known as 'lazy beds' for growing potatoes. In many parts of the Aran Islands it thus becomes possible, by putting down alternate layers of sand and seaweed, to raise a potato crop where without seaweed none could survive. We have seen that the use of seaweed for making soil may be practised on poor subsistence farms. In the case of the Aran Islands, however, a wide range of crops and vegetables, including peas, parsnips, carrots, cauliflowers and cabbages, are grown on seaweed soil, and the standard of husbandry is high.

Seaweed and sand serve to make substitute soil in other parts of Ireland and Scotland, including the Scottish islands. Potatoes are the main crop grown, but turnips, swedes, oats, etc. are also raised.

Much farther north, at some of the northern posts of Hudson's Bay

Company, vegetables are grown on seaweed, fish waste and sand. The *Polar Record* reported in 1943 that vegetable production had been particularly successful at posts on the Lower Mackenzie. Where there was no soil, a garden could be made from two 6-in. layers of seaweed, with a 2-in. layer of fish waste or garbage between, the whole covered with 2 in. of sand. If the heap was kept moist until the seaweed had rotted, and then dug up and mixed, it could be used for growing vegetables.

Among areas where high-yielding soil is made richer still by the addition of seaweed are Cornwall, Thanet, Ayrshire, East Lothian and the Channel Islands. Market garden and arable crops grown in these areas are valuable enough to make the effort of carting and spreading seaweed worth while. Treatments vary. In Cornwall, where in coastal districts seaweed has been composted with straw, Tamar Valley growers use it for their strawberries. In the Gulval area it has also been put on potatoes. In Thanet, where farmers believe seaweed must be ploughed in within twenty-four hours of gathering to get best results, harvesting and spreading are large-scale business. Contractors there use men and plant to bring up cast weed from the beaches in winter, delivering it at farms for between 30s. and 40s. a ton. That is why it may be difficult, in some seasons, to sell seaweed meal in Thanet.

In Ayrshire, freshly gathered weed is spread over the ground in autumn, the prime object being to fertilize the early potato crop. It is often spread to depths of 6–9 in., and left to rot on the surface before being ploughed in. In East Lothian, where it is used in particular on the fertile land at the back of Dunbar, it is stacked in autumn, and left to rot over the winter. So valuable is this harvest of the sea in East Lothian, that when in 1959 it was proposed to set up near Dunbar a factory to make industrial alginates from local seaweed, farmers protested that all that was available was needed for farm manure: so the factory plan was dropped.

In Jersey, seaweed is used straight from the beach, or after being partly composted. If spread on grassland it is left to rot; if spread on arable it is ploughed in. Farmers there pay more for broken-down seaweed than for fresh; but whatever the form in which they use it, the Jersey farmers are well aware of its value. I have seen a field of about 5 acres in Jersey on which potatoes have been grown without a break for ninety-two years—and, fertilized with seaweed, again without a break for the same period.

Farmers in Brittany share this respect for seaweed, and some claim that they can grow onions in the same field year after year because they give constant applications of seaweed. Some Breton farmers also put individual seed potatoes intended for the earliest crop of all on their own separate heaps of seaweed before covering them with soil. Fresh seaweed is often used in Brittany as a dressing for asparagus beds. There is also in this corner of France a considerable trade in a specialized type of seaweed (or rather seaweed deposit) known as 'maerl'. The word maerl presumably comes from the same root as 'marl'. Marl is a form of chalky clay found inland in England and elsewhere which has been used as a fertilizer for centuries. But although in function and composition there are similarities between maerl and marl, they are in origin two quite different substances.

Maerl is formed by a few red seaweeds which accumulate large quantities of calcium and other minerals to form encrustations of moderate size. These seaweeds are always submerged, and sometimes their growth is dense. When this is the case, a vast mineral deposit occurs. Whole beaches, such as the white sands, or white coral beaches, of Connemara may be formed from these maerl deposits.

Maerl is a mixture of 80–85 per cent calcium carbonate and 10–15 per cent magnesium carbonate, with about 0·5 per cent iron, and traces of zinc, iodine, boron, cobalt, copper, manganese, and other elements. Like other lime-bearing minerals, these deposits have been used in agriculture on a large scale for centuries. Even so, fresh quantities are constantly being washed up, and it is evident that extensive supplies exist in off-shore waters.

The recorded history of the use of these deposits in France dates from the twelfth century. Something like 200,000 tons of maerl are used every year in Brittany, some on the land and some, sun-dried and ground, in animal feedingstuffs. In this last connection it is valued for its balanced calcium and magnesium carbonate, and is believed to prevent tuberculosis and foot-and-mouth disease. When imported from St. Malo for use as lime in Guernsey, maerl was fetching £4 10s. a ton in December 1964.

Maerl was formerly used in parts of Scotland. The ruins of a lime kiln used for making quicklime from maerl can still be seen on the island of Bute.

On the Ile de Ré, off La Rochelle on the Bay of Biscay, cast seaweed is put on barley fields before sowing, and again when the seedlings appear; this procedure is said by Dr. V. J. Chapman, Professor of Botany at Auckland University, in his book *Seaweeds and their Uses* (Methuen, London, 1950) to have made crop rotation unnecessary.

He gives an indication of the value placed on seaweed in the 'golden belt' (the seaweed-manured coastal lands of Brittany and Normandy) by quoting a popular saying of the region: 'Point de vraic, point de hangard'—no wrack, no corn; and he also quotes an ancient statement on the use of seaweed in the Channel Islands which is worth repeating in full. 'The winter Vraicq being spread thin on the green turf and afterwards buried in the furrows by the plough, 'tis incredible how with its fat unctuous substance it ameliorates the ground, imbibing itself into it, softening the clod, and keeping the root of the corn moist during the most parching heats of summer.' That was Falla, writing of Channel Islands agriculture in 1694.

I believe seaweed fresh from the sea has also been used in various parts of Europe after being composted with hen droppings. These droppings are wet and dense, so that aerobic bacteria cannot easily penetrate the mass to carry out their work of decomposition. Composting is therefore made easier if leaves, or straw, or seaweed, are mixed with the droppings to help penetration of bacteria (and also lower that concentration of nitrogen which sometimes makes untreated droppings unsuitable for manuring certain crops).

As we have already seen, the seaweeds commonly used on farms in these islands are brown seaweeds, members of the Fucaceae and Laminariaceae families. They have the same nitrogen content as farmyard manure, and are richer in potash but poorer in phosphate. This is demonstrated in the following table.

TABLE 5

Farmyard manure and seaweed

| | *Organic matter in ton of farmyard manure* | *Organic matter in ton of undried seaweed fresh from sea* |
|---|---|---|
| | lb. | lb. |
| Organic matter | 380 | 400 |
| Nitrogen | 11 | 11 |
| Phosphate | 6 | 2 |
| Potash | 15 | 27 |

The fact that seaweed, the equal of farmyard manure in nitrogen, is rich in potash although poor in phosphate, makes it an excellent manure for mangolds, sugar beet, clover, cabbages and grass—and, of course, potatoes. The higher common salt content of the Laminarias is also of advantage with crops such as sugar beet and asparagus. If adverse effects are feared from this salt content—and I know of none—they can be guarded against by applying seaweed in winter. In any case *Ascophyllum nodosum*, more commonly used for direct fertilizing, contains less than 5 per cent of salt.

There have been very few trials to compare fresh seaweed with farmyard manure, although in Ireland, in 1929, dressings of 15 tons of seaweed to the acre produced a slightly less heavy crop than a similar dressing of farmyard manure. It should, however, be borne in mind that fresh seaweed has certain advantages which may not be reflected in yield in a one-year potato trial—advantages which may, in fact, take more than one year to become apparent. We will deal with them in greater detail when we come to consider the use of processed seaweed, both meal and liquid extract. But it can be said here that fresh seaweed, which shares all the qualities of processed seaweed, has a wider range of trace elements than farmyard manure; can improve the mechanical condition of the soil because of attributes not possessed by manure; and is always free of weed seeds and spores. Practical experience has also shown that fresh seaweed is of outstanding value on any sandy soil, particularly one deficient in potash; and that barley, in particular, usually grows well on light land to which seaweed has been added.

Subject to what has already been said about the partial composting of seaweed in Jersey, it is generally accepted that unprocessed seaweed should be put on the land in its natural state. It is also thought that if full value is to be derived from it, the sooner it is applied after gathering, the better. On the other hand, fresh application means seasonal use: it is the need to conserve seaweed for periods when it cannot easily be gathered, but may be most needed on the land, that has led to attempts at composting in the Isle of Man, Cornwall, the Scilly Isles, and elsewhere.

In Cornwall, layers of seaweed have been composted between layers of sea sand and farmyard manure and, in Scilly, between layers of soil. I find it difficult to believe that these methods are widely used now. As recently as 1941, however, the use of seaweed as bedding for cattle was advocated in the Isle of Man, and it would not be stretching terms too far to regard the resulting material as composted seaweed.

In Canada and Denmark, seaweed has been stacked on top of straw, peat, or farmyard manure, so that its valuable constituents could be absorbed as they wash out. Whether this promising technique has been tried here I do not know.

Sometimes seaweed is burnt and its ashes are put on the soil. This has been done in the Channel Islands, but I imagine it to be of rare occurrence elsewhere. Direct application is the rule; and this, in view of the bulk of seaweed, weighed down with its water content, must mean direct application within a short haul of the sea. Spreading, with or without ploughing, is the general practice. The only other we have come across, apart from those earlier described, is that of using seaweed as a top dressing in the bottom of drills after root seed has been sown. This conserves moisture, but hinders cultivation until the seaweed has decayed.

So much for historical, and present, uses of natural seaweed in agriculture. The historical use of processed seaweed in industry is further from our central subject, but it may be dealt with briefly in outline.

Until this century, apart from the product of one plant at Boothby in Connecticut which made seaweed meal in the 1870's for fertilizing tobacco plants, all processed seaweed went to industry. It was first used for making soda and potash for glass and soap manufacture; and then, since seaweeds are the richest known source of organic iodine in the world, for making iodine. Both processes are now obsolete.

Before soda, potash or iodine could be derived from seaweed, the raw material had to be burnt (as it was burnt, incidentally, to produce common salt in the Middle Ages). Burning was initiated on a commercial scale in Ireland, Scotland and elsewhere in Europe, in the eighteenth century, and the procedure was carried on until well after the Napoleonic Wars. Sporadic burning also took place until 1932 in the Orkneys, and until 1950 in France. Seaweed was also used as one source of iodine in Glasgow until 1934.

Two boom periods were the War of American Independence (1775–83) and the Peninsular War (1808–14). Three things are said to have put an end to the employment of seaweed as a raw material in the production of soda and potash: the introduction of Keir's caustic soda process in about 1800; the commercial application of Le Blanc's soda process, discovered in 1791 but not exploited until 1823; and the discovery of mineral potash in Germany. (Le Blanc's process, since replaced by other methods of synthesis, was a way of producing soda from salt with the aid of sulphuric acid, coke and limestone.)

Iodine as a separate chemical element was discovered by a chemist in Paris in 1811. He came across it in the solution of burnt seaweed in water then used in making soda, and six centres were set up in France during the Napoleonic Wars to exploit his discovery. Seaweed-growing areas which had suffered from the introduction of the industrial method of producing potash now had a second period of prosperity, and burning for iodine began in all producing areas—Ireland, Scotland, Norway, the Channel Islands and France.

The actual burning, both for iodine and potash, was done in trenches or, in some cases, in primitive kilns made of stones. The remains of some of these kilns can still be seen on the west coasts of Scotland and Ireland.

At first iodine was extracted from burnt seaweed ash with water. In later years the seaweed was not burnt but dried in sheds, and then heated to destruction in iron retorts to give illuminating gas, heavy oils, ammonia, potash, bromine, iodine and charcoal. This saved the volatile products which were lost by burning, and increased the amount of iodine obtained; but the technique proved too costly.

The discovery in Chile in 1840 of nitrate deposits, from which iodine can be more economically processed, led ultimately to the rejection of seaweed as a source of iodine. Reports from Russia in 1936 referred, however, to a process for obtaining iodine from seaweed by electrolysis. Basically, the method was to treat a solution made from macerated seaweed with an electrical charge which, as its power was increased, produced first iodine, then bromine, and finally chlorine at the anode, with cellulose and alginates at the cathode: an interesting technique, but of little commercial significance at the moment.

It will be seen that the first chemicals to be processed from seaweed were all inorganic—that is, soda, potash and iodine. In more recent times organic chemicals have been sought in seaweed—the highly complex gelatinous compounds known as alginates. This second development resulted from a discovery made by E. C. Stanford when he was working in the west of Scotland; he found that many brown seaweeds, particularly of the Laminariaceae, contained a viscous substance which, when treated with sodium carbonate and then with a mineral acid, produced a new compound which he called alginic acid. Stanford's discovery of alginic acid was first described in British Patent 142 of 1881, and information on the process was given in scientific journals in 1883 and 1884. Commercial development of his discovery, and of all that it implied, did not take place until 1934, when a newly formed company began making alginates for use in

1a. Caging black bean aphides in preference cages on seaweed-treated and untreated plants.

1b. Close-up of preference cage showing aphides on seaweed-treated plant (*right*) and on untreated plant (*left*).

2a. Right-hand sweetpea seedling raised in compost without any base fertilizer. Fibre pot soaked in seaweed extract for 24 hours and then dried. Left-hand seedling in untreated pot and John Innes seed compost.

2b. The root systems of the two sweetpea seedlings.

food and industry from the alginic acid in seaweed. This was the beginning of the alginate industry in this country. I shall deal further with alginates and agar in the next chapter.

So much for seaweed processing. Now a word about its harvesting. Seaweed, whatever its final use, has been harvested in much the same way for centuries, either by taking it off the beach where the tide has left it, or by cutting it off rocks on which it grows. In more recent years attempts have been made, some successfully and others not, to perfect mechanical underwater harvesting. But of all these methods, that of cutting seaweed off the rocks with a sickle is the one most used today. All the seaweed my company uses is harvested in this way.

If seaweed is cut on beaches inaccessible to ordinary transport, it may be formed into a floating mattress for easier harvesting. Nets may be used for its collection, or a rope may be laid out on the rocks in a circle 30–40 ft. in diameter. Seaweed inside this circle is cut, and left lying. As the tide comes in, the seaweed and the rope surrounding it both float. The mattress of floating weed can then be towed by boat to a more accessible beach for collection by lorry at low tide. In some cases it is possible for several of these rafts of seaweed to be towed away at once. This technique makes possible the harvesting of large amounts of seaweed otherwise inaccessible except by boat; and to harvest seaweed by loading it into a boat is an expensive business.

A number of mechanical seaweed harvesters have been designed for use in this country. From a commercial point of view, none has functioned successfully on our irregular and rocky coasts. The only areas where mechanical harvesting has been sufficiently rewarding are the coasts of California and Tasmania, where a large weed known as *Macrocystis*, with holdfasts on the sea bed and floating fronds at the surface, is cut by gigantic underwater mowing machines attached to sea-going vessels. *Macrocystis* is the seaweed most easily harvested by machine. The possibility of growing it off the Atlantic coast of Europe was discussed after the war, but (perhaps wisely) no action was taken. There is little doubt that if spores of *Macrocystis* were introduced here they would grow; the problem is rather that like the rabbits in Australia, they might grow too well.

Seaweed harvested in the British Isles is rarely cut oftener than once every three years. This is not for the sake of the seaweed, which will continue to grow if 8 in. or 9 in. are allowed to remain, but for the sake of those who harvest it. Seaweed grows slowly. If the whole of a particular growing area is harvested yearly, a 4-in. cut may produce

1,000 tons. If only a third of the same area is harvested yearly, and each time a different third, a 12-in. cut from this third may also produce 1,000 tons. But since it takes almost as much time and trouble to harvest seaweed in 4-in. cuts as in 12-in. cuts, the 1,000 tons harvested from a third of the area can be won in perhaps a third of the time taken to win 1,000 tons from the whole area. These figures are completely arbitrary, but they show the value of cutting different plots in turn every three years or so, rather than the whole area once a year. This is the reason why farmers on Galway Bay in western Ireland allow harvested areas to remain fallow for three to five years. It is possible to see there plots of seaweed-covered coast which have been cut one, two, three, four and five years before.

Research by the Irish Institute for Industrial Research into *Ascophyllum* beds on Galway Bay has confirmed the value of this practice. The Institute found that if holdfasts and stumps were left after each cutting, and a period for re-growth allowed, there was no danger of reducing the amount of seaweed available for later harvesting, even if the crop were cut several times in succession. The Institute recommends a re-growth period of five years as the most profitable.

Little has been done to encourage the growth of seaweed artificially round our coasts, although farmers in the west of Ireland used to put stones on sandy beaches to provide anchorages where otherwise there would be none. The stones, about 1 ft. square, were laid in parallel rows so that carts could be driven down the rows and loaded direct. I doubt whether this practice continues today—although it is always dangerous to say that any practice, especially one so traditional and simple as this, has come to an end.

The time may well come for seaweed to be cultivated here as it is already cultivated in the Far East. Certainly one of our early seaweed enthusiasts—a clergyman, incidentally—suggested anchoring enormous boom-like rafts offshore to provide sheltered lagoons for the cultivation of seaweed. His idea has never been put into effect in the British Isles, although similar techniques are used in the Far East, where seaweed has been cultivated by this and similar methods for centuries.

4

# NON-AGRICULTURAL USES OF SEAWEED

Seaweed has three non-agricultural uses: as a food, in medicine and in industry. I will give brief details of each before dealing with the main subject of seaweed in agriculture and horticulture—first because there is value in a quick glance at the other uses of seaweed, so that its agricultural and horticultural uses may be seen against a wider background; and second (as far as food and medicine are concerned) because men are animals, and if seaweed is of value to them physically it may be—indeed it is—of value also to animals, and for largely the same reasons.

Various kinds of seaweed are eaten raw, or cooked, in Scotland and Ireland, but only in South Wales is there anything more than a local trade in natural seaweed for direct human consumption. The seaweed in question, *Porphyra*, is a delicate red plant which is picked, washed in water, and boiled to produce a dark brown jelly sold on market stalls and in shops. The housewife heats it in a pan, and serves it on toast with vinegar or lemon juice. It can also be made into cakes with oatmeal or fried in bacon fat (which is the most usual way of serving it). It is known as laver bread; and the chances are, since the word 'laver' is Latin for 'water plant', that it has been eaten in South Wales since Roman times. It is certainly mentioned by Camden in his *Britannia*, which was published in the sixteenth century. Swansea is the main centre for the production of laver bread, the main markets being in and near Swansea, Neath, Llanelly and Cardiff. Chapman, who describes the preparation of laver bread in some detail, tells us that it is the classical accompaniment of Welsh mutton, and that Welsh miners are, or were, its biggest eaters.

A preparation similar to laver bread, and known as 'sloke', is sold by fishmongers in Ireland, where it is eaten with potatoes and butter.

Although *Porphyra* is the only edible seaweed gathered for commercial exploitation in the British Isles, other seaweeds are used on a domestic scale in Scotland and Ireland. These include *Alaria esculenta*, or 'murlins', and *Rhodymenia palmata*, known as 'dulse' in Scotland and 'dillisk' in Ireland. This last has been eaten cooked, either as a separate vegetable or in a stew, although far and away its commonest use is for chewing raw. In times past it has served as a substitute for chewing tobacco—as has a type of *Porphyra* known in New Zealand as 'karengo', which during the last war was sent to the Middle East for Maori soldiers, who found it more thirst-quenching on marches than chewing gum. Seaweed of (to me) unknown type has also been used instead of salt for preserving cheese in Uist and St. Kilda, and as 'teething rings' for children in the Hebrides.

Young stalks of *Laminaria saccharina*, or 'sugar wrack', which contains more sugar than most seaweeds, used to be sold in the streets of Edinburgh a century ago. Chapman, who records this, and who followed the example of an earlier French algologist named Sauvageau in sampling the seaweeds he studied, has tasted these young weeds, and says he likes them immensely, as they remind him of peanuts.

One seaweed used in home cooking is *Chondrus crispus*, or 'carrageen', which is an ingredient of blancmanges, moulds and savoury milk jellies. A recipe using carrageen in this way is given by Chapman. The same seaweed has in times past also been combined with cocoa, lemon, cinnamon, and other flavours, to make drinks.

Seaweed is eaten on a much larger scale in the Far East than in these islands, particularly in Japan. Mr. K. Wada, a Japanese nurseryman with whom I correspond, sends me packets of dehydrated seaweed soup; when dissolved in hot water, these produce what is to me a delicious and refreshing soup with a real sea tang about it.

I also noticed that Reuter's news agency reported on 11th June 1966 that Japan's oldest citizen, 114-years-old Mr. Jubei Nakamura, recommended a diet of seaweed and raw fish to guarantee a long and healthy life. 'I've hardly ever eaten anything else,' said Mr. Nakamura at his birthday celebrations in Kamaishi.

The birds' nest soup esteemed a delicacy in China derives from seaweed. The swallows which make these nests and attach them to the roofs of caves build them up with successive layers of soft, gelatinous, regurgitated seaweed, which hardens in the air.

Dr. W. A. P. Black, of the Scottish Institute of Seaweed Research, writes in the journal *Agriculture* that seaweed has for centuries been an accepted food for human consumption in the Far East, up to a quarter of the people's diet there consisting of seaweed. They eat seaweed from childhood; and as a result, their intestinal microflora (or beneficent bacteria) essential for digesting certain of the seaweed constituents, are now able to extract considerable nourishment from the plant. Dr. Black suggests that if the appropriate microflora could be developed generally in men and animals, 'the whole aspect of seaweed as a food might be changed'.

Natural seaweed also has medicinal qualities. Carrageen in milk is used in Ireland for relieving chest troubles, and a diet containing the weed is recommended for building up weakly children. Cooked seaweed is said to have cured calves of 'wasting disease', while recent experiments with guinea pigs have shown that it will reduce gastric ulceration. Its dried fronds, gelatinous and easily digested, have for this reason been prescribed for invalids. Carrageen poultices are said to cure boils, while anyone with a whitlow on his hand who pushes it into a pile of rotting carrageen and leaves it there for as long as he has patience will find (or so they say in Ireland) that his whitlow is cured on withdrawal. The escape clause here is 'for as long as he has patience'.

Dulse used to be used in Skye to induce sweating during fever. Sauvageau tested its qualities and, according to Chapman, 'found no cause for dissatisfaction'. In Iceland, seaweed has been used for over a thousand years as food for man and beast, as well as for fuel and manure, and for providing dyes for homespun materials—as it has, incidentally, in the island of Harris. It has also been used in Iceland to protect plants from frost, as a filling for cushions, and for primitive lampwicks. *Rhodymenia palmata* has been taken as a remedy for seasickness, as a substitute for vegetables, and as a source of vitamins. *Alaria esculenta* is also cooked and eaten there.

A more certainly effective use of seaweed in simple medicine is for keeping open wounds which need to be drained. Strips of dried *Laminaria* used to be put into wounds to absorb the discharge. In swelling up, these strips prevented closing of the wound and so helped it to continue draining.

I have now dealt, in rather cursory fashion, with the use of natural seaweed in what might be called home cooking and nursing. Before dealing with the use of seaweed extracts in food and in industry, I must first say something about the use of seaweed from the point of view of modern medicine.

## *Non-agricultural uses of seaweed*

Seaweed meal is a simple product—whole seaweed dried and ground, no more, no less. The seaweed extracts used in food, medicine and industry are, however, complex products, and their manufacture requires some explanation. For this reason I shall first dispose of seaweed meal, and then deal with the more complex matter of processed seaweed extracts.

It is mainly the mineral trace elements and vitamins in seaweed which make it so valuable an addition to the diet. For this purpose it is available for human consumption in the form of a fine, dry powder which is eaten either as powder or in the form of pellets. I, and many others, take a small teaspoonful daily on porridge or breakfast cereal, or in milk, as naturally as one would eat fruit or greenstuff. It is best taken in small amounts to make it more easily digested; in theory, at least, the starches and sugars contained in the commoner seaweeds are not easily digested in their untreated state by human beings. There is also much evidence to suggest that seaweed is more effective in small doses as food for men, animals and plants.

It is generally accepted that natural seaweed can offer valuable relief in rheumatism—the hydrolized extract we make for horticultural use is added to hot baths to give some degree of ease in this complaint. Seaweed is also a useful source of organic iodine for those who need it. The orthodox would probably admit no more. But there is evidence, of varying worth, that seaweed meal can be of value in treating influenza, mumps and hay fever, and I give this evidence for what it is worth.

Researchers in New Jersey who infected chicken embryos with influenza and mumps virus to test the effectiveness of various drugs, found that they were getting misleading results because the seaweed extract in which the viruses were held tended to retard their growth. It has also been suggested by workers at the Scripps Institute of Oceanography in the University of California that seaweed may inhibit hay fever. This complaint is comparatively rare in Japan and China, where seaweed is commonly eaten; but immigrants from those countries are said to have developed it after settling in the United States, where seaweed no longer formed part of their diet.

Some believers in the curative properties of seaweed—or kelp, as it is often called in this connection—take things further than this. A nostrum known as 'Anti-Fat', a remedy for obesity marketed in the 1880's, was based on seaweed, while I know of at least one beauty consultant who adds a pint of our liquid seaweed extract to each bath of hot water she gives her charges—to encourage slimming.

I also have before me a recent publication which recommends the use of seaweed against exophthalmic goitre, obesity, excessive thinness, nervous complaints, deficiency diseases, nervous dyspepsia, high blood pressure, constipation, biliousness, and tension and pain in the back of the neck and the base of the skull. The author claims that seaweed has a beneficial effect on the gall bladder, pancreas, bile duct, kidneys, prostate gland, uterus, testicles, ovaries and thyroid gland. Perhaps I should explain, for the serious minded, that this passage is quoted for light relief. I think that seaweed is of value, but I would not take the claims quite so far. My own belief is that a small amount of seaweed, eaten in pulverized form over a long period, does help to improve one's general physical condition, although (to quote from the work mentioned above) 'it is one of those slow persistent agents that require time to accomplish the desired results'. Such details as I have given of the traditional uses of seaweed in 'folk' medicine, as well as more modern pointers to the possible value of seaweed in other directions, all suggest that there is much scope for further research into its curative properties.

I now turn to the use of seaweed extracts, and here it becomes necessary to describe two of them in some detail: agar and alginates.

The name 'agar' is derived from the Malayan word agar-agar, whose double structure indicates plurality. The word was first applied, in the Far East, to a kind of red seaweed used for making jelly, and then to the jelly itself. It is made from a limited group of seaweeds, of which *Gelidium* is the most important. Japan has always been the world's main producer of agar, but when supplies from that country were denied us during the war, we produced a substitute for it from *Gigartina stellata.*

Agar is a simple gelatinous extract of seaweed, usually made by concentrating the solution which results from an infusion of seaweed in hot or boiling water. It is not one specific constituent of seaweed, of known chemical composition, which is divided off from the rest—as, for example, the complex chemical sugar is divided off from beet or cane to provide a standard product. It is, on the other hand, a product whose precise chemical composition may vary within certain limits. The use of the word 'agar' to describe this simple gelatinous extract is governed by association, as well as by inherent characteristics. The product is known as 'agar', for example, when used in medicine; but a similar gelatinous extract of seaweed may not generally be known as 'agar' when used, for example, to provide a bed for tinned mackerel.

An attempt to define the best quality Japanese agar has been made. An American research worker describes it thus: 'the dried, amorphous, gelatine-like non-nitrogenous extract from *Gelidium* and other agarophytes (his word for seaweeds from which agar is made), being the sulphuric acid ester of a linear galactan, insoluble in cold, but soluble in hot, water, a one per cent solution of which sets at 35 to 50 degrees centigrade to a firm gel, melting at from 80 to 100'.

Recent work has modified this chemical definition. The distinguishing chemical feature of agar has now been found to be the presence of a high proportion of 3, 6-anhydro galactose units; the sulphate ester is unimportant. But speaking more generally, the most useful distinguishing feature of agar is that it can form a gel setting at from 35° to 50° C., but melting only when heated to between 80° and 100° C.

If agar is described as an extract obtained from seaweed by physical means, then alginates are extracts which are obtained by chemical means. Yet although alginates are specific chemicals separated from seaweed by a controlled chemical process they can, and do, have many of the characteristics of agar, of other seaweed products, and of seaweed itself.

Alginates are the salts of alginic acid, and their name comes from the Latin word for seaweed, algae. Alginic acid makes up from 10 to 30 per cent of the dry matter of brown seaweeds only, as the structural material of the cell walls. Like all living tissue it is made up of the three elements of carbon, hydrogen and oxygen; although unlike some living tissues, purified alginic acid contains no nitrogen. Alginic acid is the main structural carbohydrate of seaweed, just as cellulose is the structural carbohydrate of land plants.

Stanford (*see page* 48) discovered alginic acid as a result of treating seaweed with hydrochloric acid and sodium carbonate; and it was this discovery which, after a considerable lapse of time, brought about a complete change of attitude to seaweed on the part of the chemical industry. Whereas at one time seaweed was regarded as an important source of inorganic chemicals, in particular of iodine and potash, it then became an even more important source, for new industries, of organic chemicals—the organic colloids we know as alginates.

Although agar and alginates are two separate products, their uses and applications have tended in recent years to overlap. The first use of agar, at least in the East, was as a gelling agent for food. Its use as a base on which fungi, bacteria and other microflora could be cultivated was a later development, which began in Europe and spread throughout the world. It is still used as a gelling agent in food, and as a laxative.

In practice, however, most of the uses to which agar has been put in food and medicine are now discharged by alginates; and alginates have found a number of entirely new uses in industry where agar has not been tried. There is no need for us to detail the former uses of agar; it is enough to remember that some of the applications of alginates in food and medicine may at one time—may still—be performed also by agar. Let it also be remembered that the two terms, although not synonymous, are at times used as if they were.

The tendency for confusion to arise over 'agar' and 'alginates' is made worse by the existence of a third, comparable, seaweed product: carrageenan. Carrageenan, like agar, is extracted from red seaweeds; in this it differs from alginates, which are made from brown seaweeds.

Carrageenan, and carrageenan-type products, are obtained from a large number of red seaweeds, of which Irish moss (*Chondrus crispus*) and *Gigartina stellata* are the most important. Chemically, carrageenan is somewhat similar to agar, although it contains a high proportion of ester sulphate. The so-called British agar manufactured during the war was made by the chemical processing of a *Gigartina* extract (that is, carrageenan). In spite of its name, it did not have quite the same characteristics as natural agar.

Extracts of the carrageenan type are now probably used as much as agar in the food and other industries. They are used for the formation of weaker gels than that produced by agar—in milk puddings, for example. and for thickening and stabilizing. Much carrageenan, particularly in America, is used for stabilizing the cocoa used in making chocolate, in milk drinks, and in toothpastes.

The production of agar is about twice that of carrageenan, and its price is approximately double.

There are, in fact, a number of red seaweed extracts on the market which sell under different names. Although of slightly different chemical composition and slightly different properties, they can all in practice be regarded as of agar, or of carrageenan, type.

To return to alginates. Here is a table showing some of their main uses.

TABLE 6
Some uses of alginates

| *Function* | *Industry* | *Product* |
|---|---|---|
| Thickening | Food and bakery | Filling creams<br>Lemon curd<br>Soups |
| | Pharmaceutical and cosmetic | Tooth paste<br>Hand creams<br>Shampoos<br>Lotions<br>Liquid detergents |
| | Textile | Printing pastes |
| | Rubber | Latex |
| Emulsifying, stabilizing and deflocculating | Food and bakery | Ice cream<br>Soft drinks<br>Bakers' emulsions |
| | Pharmaceutical | Emulsions |
| | General | Polishes<br>Emulsion paints<br>Welding electrodes<br>Insecticides |
| Gelling and binding | Food and bakery | Confectionery jellies<br>Packing jellies<br>Milk jellies and puddings |
| | Pharmaceutical and cosmetic | Tablets<br>Hand and anti-burn jellies |
| | Dental | Dental impression powder |
| | Medical | Haemostatics |
| | Ceramics | Glazes and engobes |
| | General | Sintered products |
| Film and filament forming | Food | Sausage casings |
| | Pharmaceutical | Barrier creams |
| | Medical | Absorbable dressings |
| | Textile | Soluble yarn<br>Warp sizes |
| | Paper | Transparent paper<br>Coated papers<br>Washable wall papers |
| | General | Anti-stick and mould release agents<br>Leather |

*Note:* Alginates are used with welding electrodes as part of the flux, and with sintered products as a temporary binding agent. Engobes are a form of pottery decoration; warp sizes are used in the textile industry to get better weaving performance.

It may be guessed that processed seaweed in the form of alginates is used on a greater scale in the food manufacturing industry than is natural seaweed in home cooking. Alginates are used to make liquids viscous, to produce and stiffen jellies and blancmanges, to make emulsions, and to thicken soups. They are used in the manufacture of what are elliptically termed 'bakers' sundries'. They are used to stabilize milk used in manufacturing processes, and so prevent it separating on standing; and in the fining, or clearing, of beer and other liquids. Alginates make it easier to whip ice-cream and give it a smooth texture; they also prevent the formation of large crystals during its storage, and of curding when it melts. Indeed, so important is seaweed to the ice-cream industry, that according to one authority, half the total production of alginates in the United States went at one time into ice-cream. Other uses of alginates have, however, developed faster than this, and now the amount so used is perhaps between 10 and 20 per cent of total United States production.

Synthetic cream, fruit squashes and other drinks can contain alginates as stabilizing agents. Sausage skins can be made of alginates, while the jelly in which tinned fish is often set to prevent the soft, cooked tissues breaking up in transport can be made, wholly or in part, of alginates. The jelly in meat pies may be stiffened with alginates, which also find other uses for salad dressing, soups, confectionary, meringues, and so on. Alginates are employed, pharmaceutically and cosmetically, as thickeners in toothpaste, hand creams, cough mixtures and lotions, as a base for ointment, and as a substitute for glycerine.

Alginates—and in this particular case, agar—have been used to give bulk to invalid foods such as those prepared for diabetics, assuaging hunger without taxing the system. Their bulk and bland nature have also encouraged their use as a laxative, stimulating intestinal action without causing irritation which chemical agents might occasion.

In medicine, gauze or wool spun from calcium alginate filament arrests bleeding; and since it is absorbed by the body, can be used to arrest haemorrhage at points from which, once applied, it could not be removed. Burns can be treated by covering the whole of the damaged area with a protective film of sodium alginate gauze.

The fact that alginates can mould themselves gently to uneven surfaces, and remain in that mould although momentarily distorted in removal, has led to their being used for taking impressions of parts of the body, notably teeth, eyes and hands. Alginates are used in dentistry, in particular, to take an impression so strong and elastic, that when set it provides an accurate reproduction even of undercut cavities.

For the future, the role of seaweed in preventive medicine may well be completely revolutionized by the recent discovery at McGill University, Toronto, that sodium alginate reduces the absorption of radio-active strontium in food without the harmful side effect of interfering with the absorption of calcium also. This finding is elaborated in some technical detail by Stanley C. Skoryna, D. Waldron-Edward and T. M. Paul of the Gastro-Intestinal Research Laboratory at McGill, in a paper published in the *Proceedings of the Fifth International Seaweed Symposium* (Pergamon Press, Headington Hill Hall, Oxford).

In textile manufacture, calcium alginate is used to form temporary supporting or connecting thread—scaffolding thread, as it is known in the trade—which, since it is soluble in soap and soda solution, can be easily removed when the garment is washed during finishing. The same thread can be used as a foundation for embroidery, so that when washed away it leaves a net-like effect similar to lace, or a series of holes at desired intervals. It can be used to make worsted fabrics weighing as little as 1½ oz. a square yard, by being woven with yarns too fine to be woven on their own. When the resulting material is washed, the scaffolding thread disappears, leaving a fabric far lighter than could otherwise be made.

The use of calcium alginate as scaffolding thread has, however, become of lesser importance to the textile industry than was at one time anticipated. Its main use is now in the knitting of socks, making unnecessary the separation, by cutting, of a continuous series. When the alginate thread used to connect the socks disintegrates on washing, the socks fall apart from each other automatically.

A more important textile use for alginates is as a thickening agent for the dye used in printing—sodium alginate is the only one of a number of such agents which can be used with the fibre-reactive dyes now used.

In the soap industry, alginates are used to thicken liquid detergents. In papermaking, alginates are used as surface sizing to improve the printing quality of paper. They are also used—or were used during the war—to make soluble paper for secret documents, which could be destroyed by contact with water. Finally, to add two more unusual uses for seaweed products, let me mention that of perfume-making in France and of board-making in Ireland.

Those who manufacture alginates say they have a thousand industrial uses. I have no doubt they are right.

5

# HOW PLANTS GROW

A knowledge of the soil, and of how plants grow, will help us to understand why seaweed is of value in farming and horticulture.

The description we now give of the characteristics and functions of plants and soil may seem, on first reading, to depart too far from our main subject of seaweed in agriculture. But in fact nearly all the matters now described—the structure of the soil, photosynthesis, the nature of the chlorophyll molecule, mineral salts in plants, enzymes and trace elements—are immediately related to seaweed, and to the use of seaweed. For this reason we have thought it better to explain these things first in general terms, as simply and as briefly as possible, so that when the time comes to relate seaweed to soil structure, photosynthesis, chlorophyll, and so on, their relationships may be more easily understood.

Soil consists of some, or all, of the following: particles of sand, silt, or clay; humus; water; air; dissolved salts; bacteria, fungi, and other minute soil organisms.

Sand, silt and clay are residues of rocks which have been broken down by weathering and other mechanisms. Of the three, clay has the smallest particles, silt the next largest, sand the largest of all. These particles, usually produced from the rocks of the district, form what might be called the skeleton of the soil, and provide the physical foundations on which plants are supported. The colour, chemical composition, and to some extent size, of these particles are often also determined by local rocks—although the overall colour of soil may be varied by its organic content, and water and ice may carry soils into areas far from the rocks which produced them.

In good growing soil, most of the inorganic particles are covered with a film of water which is held in place by capillary action; between the particles or groups of particles so covered are air spaces which provide oxygen for soil bacteria and plant roots.

In clay soils, which are made up of small and closely packed particles, the films of water round each particle exert a stronger surface tension than in soils with more widely spaced particles. For this reason clays are difficult to drain, and heavy and sticky to work with plough or spade. The smallness of the air spaces between the particles also means poor aeration of the soil, and this in turn slows up the activity of soil bacteria. On the other hand, clay soils retain moisture better in drought, and there is less danger of the mineral salts they contain being washed away by rainwater—or leaching, as it is called by farmers.

In light, sandy soils, the large inorganic particles give better aeration and drainage. Surface tension of soil water is less powerful, so that ploughing and digging are easier. But such soils are more liable to lose mineral salts and other nutrients through leaching; and they dry out rapidly in drought.

Many soils, of course, are based on particles of intermediate or of varying size. Most good soils, too, have what is called a crumb structure, or are given such a structure by careful cultivating. This means that they are made up of a mass of crumb-sized particles, which allow both air and water to be held in reasonable amounts in the soil.

Humus, another of the constituents of soil, is decaying plant or animal matter. When broken down by bacteria—and this is what decay implies—humus provides mineral salts essential for plant growth. (Further mineral salts may also be dissolved out of surrounding rock.) Because its density is low, and the air spaces in it are large, humus makes heavy soil lighter and easier to work. Its spongy qualities also help lighter soils to hold water longer than would otherwise be the case.

Those unfamiliar with the principles of plant growth might suppose that it was the soil which gave plants all the food they needed, and that water simply made this food available to plant roots. Nothing could be further from the truth. More than half the weight of plants is made up of water, which obviously comes largely from the soil; but of the dry matter which remains, 90 per cent comes not from the soil but from the air. This means that only about 10 per cent of the dry matter of the average plant comes from the soil. This was suspected over three hundred years ago by a Dutch researcher, van Helmont, who proved by experiment that plants do not take their substance from the soil. He cut a willow shoot, and found that it weighed 5 lb. He then weighed out 200 lb. of soil, and planted the shoot in it. Five years later the soil still weighed 200 lb., less 2 oz. Nothing but water had been given to the willow shoot, yet the tree which grew from it now weighed just over 169 lb. Where had the extra 165 lb. come from?

It came, as we now know, from the air, by what is called photosynthesis. Photosynthesis is the process by which plants take carbon out of the air and, by using the energy of light, turn it into materials for building up their own tissues. Photosynthesis, which literally means building up by light, is thus basic not only to the growth of plants, but also to life itself on this planet.

The reader may wonder where the carbon in the air comes from. It appears there, combined with oxygen, as the inert gas carbon dioxide, which is breathed out by animals and men as a waste product after breathing in oxygen. Plants 'breathe in' this carbon dioxide through their leaves; break it down using energy from the light; absorb the carbon; and then liberate the oxygen into the air, so that it can be used again to sustain the life of animals and men.

Once they have taken carbon out of the atmosphere, plants combine it with water—or with the hydrogen and oxygen molecules which together make up water—and in this way make entirely new substances called carbohydrates. Carbohydrates are the basis of all living tissue, both animal and vegetable. They are always composed of a number of carbon, oxygen and hydrogen molecules; but however many there may be, the number of hydrogen molecules is always double the number of oxygen molecules. This is the same as in water, which explains why the word is carbo-*hydrate*, 'hydra' being Greek for water. In water, however, there are never more than two molecules of hydrogen and one of oxygen in combination, whereas in carbohydrates the numbers may be greater. In glucose, for example, in addition to the six atoms of carbon, there are twelve of hydrogen and six of oxygen.

One other essential factor in photosynthesis remains to be mentioned: the whole process takes place in the plant's leaves, and depends on the presence there of a complex chemical known as chlorophyll, the green stuff of plants. Overleaf is a plan of one molecule of chlorophyll—the smallest amount of chlorophyll which still remains chlorophyll. Anything smaller would be a collection of loose carbon, hydrogen, and other atoms.

In the centre of this molecule, among the 136 other atoms you will see one single atom of the mineral magnesium (Mg.). This is only one example of how minerals from the soil are essential to the growing plant, even in amounts so small as this.

TABLE 7

Plan of chlorophyll molecule

```
            CH2
            ‖
            CH      CH3
            |       |
            C   CH  C
           / \\ / \\ / \
     H3C-C    C    C    C-CH2CH3
          \\  |    |   /
           C-N    N-C
          /    \  /    \
        HC      Mg      CH
          \            /
           C-N    N-C
          /   ‖   |    \
    H3CHC    C    C    C-CH3
          \ /  \ // \ //
           CH    C    C
           |     |    |
           CH2   CH-C=O
           |     |
           CH2   C—O-CH3
           |     ‖
         O=C     O
           |
           O
           |
           CH2
           |
           CH
           ‖
      H3C-C
           |
           CH2
           |
           CH2
           |
           CH2
           |
      H3C-CH
           |
           CH2
           |
           CH2
           |
           CH2
           |
      H3C-CH
           |
           CH2
           |
           CH2
           |
           CH2
           |
      H3C-CH
           |
           CH3
```

Chlorophyll takes carbon dioxide from the air and water from the plant roots and produces a kind of sugar called glucose. At the same time it liberates oxygen, as the following chemical equation shows. (C stands for carbon, O for oxygen, H for hydrogen.)

$$\underset{\text{carbon dioxide}}{6CO_2} + \underset{\text{water}}{6H_2O} = \underset{\text{glucose}}{C_6H_{12}O_6} + \underset{\text{oxygen}}{6O_2}$$

With this sugar, and the salts the plant takes from the soil, green plants can make any of the substances they need for living.

I said earlier that more than half the gross weight of most plants is

3a. Storm-cast *Laminaria* seaweed with dulse and other seaweed growing on the stipe (stalk). By courtesy of Institute of Seaweed Research, Inveresk.

3b. Harvesting *Ascophyllum* seaweed at Smola, Norway. By courtesy of Norwegian Institute of Seaweed Research, Trondheim.

4a. The effect of compost soaked in seaweed solution (*left*) on *Berberis darwinii* cuttings.

4b. The effect of compost soaked in seaweed solution (*left*) on dwarf azalea cuttings.

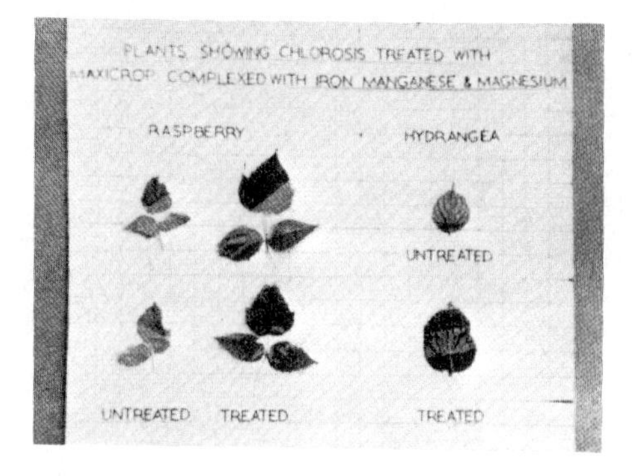


4c. Seaweed extract with chelated iron, manganese and magnesium helps plants suffering from severe chlorosis.

water; and we have also seen that water is used by the plant in making the dry matter of its tissues. Yet of the water which passes through a plant from roots to leaves, and so by evaporation into the atmosphere, only a fraction is used in making carbohydrates. This constant movement of water through the plant has other functions. The water carries dissolved salts from soil to leaves; it evaporates at the leaf surfaces, so that more water is drawn from veins to leaf, stem to veins, stalk to stem, roots to stem, and helps to keep the leaves cool in the direct rays of the sun. These are positive effects. But it does seem that some of the transpiration which takes place in plants is in a sense accidental. A large expanse of moist leaf surface exposed to the air provides an excellent way of absorbing the carbon dioxide used in photosynthesis, but it is an equally efficient way of evaporating water. This means that although the amount of water essential to photosynthesis is small, the process can be carried on only with the inevitable loss of a large amount of water through evaporation. (It is said that as much water is absorbed, and evaporated, by plants in temperate countries as runs down to the sea in rivers—which helps us to understand why rainfall failed disastrously in areas such as Arabia and parts of North Africa when they were denuded of trees and other plant life.)

The movement of water through a plant, as well as from cell to cell within the plant, depends largely on what is called osmosis. Osmosis is the phenomenon whereby water will pass through a membrane when that membrane separates pure water on one side from a solution on the other. The movement, which is from the pure water side to the solution side, takes place at molecule level. By this I mean that the water does not flow freely through the membrane in liquid form—as it does, for example, through some poor quality men's raincoats now on the market—but as separate molecules of $H_2O$.

Many of the membranes in plants act in the same way as membranes used to demonstrate osmosis in the laboratory. Plant roots, for example, are simple osmotic systems. Their tissues act as semi-permeable membranes through which water diffuses into the plant from water solutions in the soil—from the side where the solutions are weak, to that where they are strong. The root system of a plant must be able to absorb this water at a high rate. That is why the root system of a grass plant, for example, is in total often hundreds of yards in length: length is necessary to provide a large area of contact between the osmotic system of the roots and the film of water round the soil particles.

We must now return to the matter of plant nutrients, and to that 10 per cent of the dry matter of plants which comes not from the atmosphere through photosynthesis, but from the soil.

The nutrients which come from the soil are mineral salts dissolved in water. They are used in much smaller proportions than the carbon, hydrogen and oxygen used to make the carbohydrates which form the bulk of plant tissue. Nevertheless they are essential to the life of the plant. To give an idea of the kind of mineral salts plants take from the soil, here are the mineral elements found in the maize plant. They came into the plant, from the soil, in the form of the soluble mineral salts earlier mentioned.

TABLE 8

Minerals in maize

| | *Percentage of total weight of plant* |
|---|---|
| Nitrogen | 1·46 |
| Phosphorus | 0·20 |
| Potassium | 0·92 |
| Calcium | 0·23 |
| Magnesium | 0·18 |
| Sulphur | 0·17 |
| Iron | 0·08 |
| Silicon | 1·17 |
| Aluminium | 0·11 |
| Chlorine | 0·14 |
| Manganese | 0·03 |

Elements such as these are essential to the life of the plant. Crops, and animals fed on them, may suffer if any are missing from the soil or, if present, unavailable. They may also suffer if, in attempting to make good a deficiency, men give plants more of an element than those plants need.

We cannot, however, look on all these mineral elements in the same way. All come from the soil—even nitrogen, although there happens to be much more nitrogen in the atmosphere than there is carbon. But apart from that, certain divisions may be made. Nitrogen, which is used on a larger scale than other mineral elements, is essential for the formation of what is known as the protoplasm of plant cells. Nitrogen, as well as potassium and phosphorus, can be given artificially to crops to increase yield, and can be given on a large scale without obvious ill-effects. This is true of none of the other mineral elements. Some distinction, therefore, can be made between such mineral elements as nitrogen, potassium and phosphorus, which are not true trace elements (elements essential, in traces, to plant growth), and other mineral elements such as boron and manganese, which are.

Before dealing with trace elements I must say a word in explanation of the rather specialized function of nitrogen among mineral elements. We have seen that nitrogen is an essential part of the protoplasm of plants. Protoplasm, the 'living matter' of plants, is found inside cells concerned with growth. As such, protoplasm is different from the cellulose which forms the walls of those cells, or of other cells which, as in the wood of a tree, may be purely structural. To make a rough-and-ready division between these two plant substances, cellulose can be said to be made from sugar, which is in turn made from the carbohydrates produced by photosynthesis, whereas protoplasm is made mostly from protein and small proportions of ordinary carbohydrates, as well as fats, mineral salts, and what are called nucleic acids. The fact that this 'living' part of the plant, the protoplasm, contains nitrogen, explains why this element is vital to plant growth.

Nitrogen, as we have seen, cannot be taken by the plant directly from the air, although the air is full of it. It must come from the soil, in solution, in the form of nitrogen salts. In a state of nature these salts are produced by soil bacteria from decaying animal and plant materials, in particular from farmyard manure, although additional nitrogen is also supplied by a process known as nitrogen fixation. Nitrogen-producing bacteria are also found in the root swellings, or nodules, of certain members of the pea family such as clover, beans and sainfoin. It is because these plants increase the nitrogen content of the soil that they are included by farmers in three- and four-year rotations. (These remarks about plants being unable to take nitrogen directly from the air will be modified when we consider foliar spraying, whereby nutrients, including nitrogen, are absorbed in solution through the leaves; but in this context they are true enough.)

The energy which is required to build up proteins from carbohydrates, nitrogen and sulphur, comes from breaking down the carbohydrates the plant has already made. In so doing the plant produces carbon dioxide and water, according to the following equation:

$$\underset{\text{glucose}}{C_6H_{12}O_6} + \underset{\text{oxygen}}{6O_2} = \underset{\text{carbon dioxide}}{6CO_2} + \underset{\text{water}}{6H_2O} + \text{calories}$$

This is a greatly simplified summary of what is in fact an infinitely complex process; and for our present purposes it is less important to understand these reactions which take place in the living plant cell, than to know that they are initiated, and controlled, by what are called enzymes. We will later consider the subject of enzymes in greater detail. At the moment we must go back to those mineral elements which are absorbed through the plant's roots.

We have made a distinction between nitrogen, potassium and phosphorus (as elements which can be given on a large scale with apparently entirely beneficial results) and what are called trace elements, which while essential in minute amounts, can be destructive on a larger scale.

Trace elements are of value to plants, animals and men in a number of ways—how many, in fact, no one knows. But there is one way in particular in which trace elements are essential to plants—and that is in forming part of the enzymes which initiate, and control, many of the vital processes of plants. You may know that some chemical reactions are hastened by the presence, and retarded by the absence, of a certain chemical which seems to take no part in the reaction, and is apparently unaffected by it. Such chemicals are known as catalysts. A similar function is carried out in living organisms by enzymes. The carbon produced in a plant by photosynthesis can be built up into fats or sugars, within the plant, inside thirty seconds. If ever those same chemical reactions took place in identical conditions outside the plant, they would do so at an infinitely slower rate. That which accelerates the reaction within the plant is one or more of a group of enzymes—complex organic catalysts produced by the plant itself. Of 650 different enzymes from plants, animals and bacteria listed in 1958, those which cause and control fermentation are the best known; but valuable as enzymes are to the brewing and other industries, most known enzymes are said to be effective in the presence of only one chemical substance.

Now the important thing from our point of view is that the enzymes which initiate and control chemical reactions in plants—reactions which are essential to the growth and production of crops—often contain inorganic components. Sometimes the complex molecule of an enzyme with dozens, or even hundreds, of atoms of carbon, oxygen, hydrogen and perhaps nitrogen, has just one atom of a metal. Without that one atom of metal in each of its molecules, the enzyme would be unable to function. Chlorophyll is not an enzyme, but it shows the same characteristic.

In one enzyme process mentioned earlier—the oxidation, or burning up, of carbohydrates to produce energy and, ultimately, protein for the host plant—the use of an enzyme makes it possible to complete oxidation without its normal accompaniment of combustion, fire or explosion. In the case of this process, and similar biological processes, the deprivation of a trace element could therefore inhibit vital biological mechanisms and cause ill-health, as a result of interfering with the enzyme concerned.

Further evidence of the value to plants of mineral trace elements has been provided by hydroponics—growing plants, in the absence of soil, by offering them food in solution. Hydroponics confirms that seven main mineral elements are indispensable for plant growth—potassium, magnesium, calcium, nitrogen, phosphorus, sulphur and iron—although some of these are in fact needed on a larger scale than is implied by the use of the word 'trace'. It has also been found, by experimenting in laboratory conditions, that certain other elements are also essential—but in their case in much smaller amounts. These are true trace elements, and they include boron, manganese, molybdenum, zinc, copper, sodium, vanadium, aluminium, and possibly others. They are necessary in concentrations of one part in ten million, or even one hundred million, of feed solution; and while essential in such minute traces, are generally poisonous if the proportion is increased.

One example of an enzyme which contains a mineral element in its molecule is nitrate reductase, which 'reduces' nitrogen to a state in which it can be absorbed by a plant's roots. Nitrate reductase contains molybdenum—one reason, and probably not the only one, why molybdenum is essential to the healthy growth of plants. Table 9 shows which enzyme mechanisms are associated with the four trace elements of copper, manganese, molybdenum and zinc. The table, and the two which follow, include substances unfamiliar to the general reader. But they do underline the biological importance of trace elements to soil bacteria, plants, animals and men, and may be of value for reference purposes.

To explain each of the terms used in detail would take us too far from our main thesis. But as an indication of the type of substances included I may say that those ending in -ase are enzymes with various functions. Oxidase (copper) brings about oxidation, while arginase (manganese) is important in the production of urea in the liver. *Neurospora* (molybdenum) is a mould. The most important use of flavin-linked oxidase (molybdenum) is in the de-toxication of alcohol. Carbonic anhydrase (zinc) is associated with the formation of red blood cells, and carboxy-peptidase (zinc) with digestion; *Aspergillus niger* (zinc) is a mould.

TABLE 9

Enzymes and trace elements

A table showing some enzymes and the trace elements they require

| *Essential trace element* | *Soil organism enzymes* | *Plant enzymes* | *Animal enzymes* |
|---|---|---|---|
| COPPER | OXIDASES IN BACTERIA | POLYPHENOL OXIDASE<br>ASCORBASE<br>LACCASE | BUTYRYL DEHYDROGENASE<br>URICASE<br>INDOPHENOL OXIDASE |
| MANGANESE | AZOTOBACTER HYDROGENASE | ENZYMIC SUGAR BREAKDOWN<br>CHLOROPHYLL SYNTHESIS | ARGINASE |
| MOLYBDENUM | NITRATE REDUCTASE (in *Neurospora*) | | FLAVIN-LINKED ALDEHYDE OXIDASE |
| ZINC | CARBOXYDASE (in *Aspergillus niger*) | CARBONIC ANHYDRASE | CARBONIC ANHYDRASE<br>CARBOXY-PEPTIDASE |

We shall return later to the matter of trace elements, and the practical results of their absence. Meanwhile we must turn to another series of materials essential to plant growth: auxins, or plant hormones. (The word 'hormone' applies to both plants and animals, 'auxins' to plants only.)

Auxins are substances which influence the activity of the living cells in plants by encouraging, and inhibiting, the extension growth and division of cells. They can be effective in concentrations of much less than one in a million. They occur naturally, and some have been made synthetically. Synthetic and natural auxins work at concentrations of between one and ten parts per million when applied artificially to the stems of plants, and at concentrations of between one and ten parts per million-million when applied in the same way to roots. Concentrations produced and applied by the plant itself may be different.

To take one example, experiments with seedling oats show that a substance at the tip can stimulate cell division, and therefore the extension of tissue, lower down the stem. This growth-promoting substance is the auxin indolyl-acetic acid. It is an important auxin, but only one of many. The fact that it causes growth can be proved easily by what is called the oat coleoptile test. In this test, a sliver of growing oat seedling, say a centimetre long, is left in an auxin solution for a given time, and then measured by the side of a similar section of oat seedling kept for the same time in distilled water. Any increase in length of the first oat section results from the auxins in the solution. This test, which can be used with natural or synthetic auxins, is sensitive enough to show the presence of one part of auxin in many million parts of water.

The main function of plant auxins is the balanced control of speed of growth. They can help to make roots grow, prevent buds forming or opening, or encourage leaves to fall in autumn. Since synthetic auxins have the same root-forming qualities as natural auxins, they are used in nurseries to propagate plants from cuttings. By delaying or preventing the formation of a separate layer at the base of the leaf stalk, they can delay leaf fall in autumn; and when sprayed on certain plants and trees in flower, can produce seedless fruits or control the number of flowers which set fruit.

Auxins, both natural and synthetic, not only stimulate the growth of plants; they can also prevent growth at one point, while stimulating it at another. Thus one or more natural auxins which discourage the growth of lateral shoots ensure dominance by the terminal bud on a branch. This growth-inhibiting quality has been exploited by the production of synthetic auxins which delay the opening of flower buds on fruit trees, and so avoid the destructive effects of late frosts; retard growth in nursery trees and bushes during transport; and inhibit sprouting in potatoes, and other vegetables, during storage.

When auxins inhibit growth, they presumably do so by overloading the machinery of growth in some way. This overloading, taken to extremes, can kill the plant. That is why synthetic auxins are also used as weedkillers—they act, not by inhibiting growth, but by monstrously accelerating it. Not only are they effective in comparatively low concentrations, but they are non-corrosive, and said by their manufacturers to be harmless to men and animals. Some are selective, and can kill weeds without damaging the crop in which they are growing. Among the most effective are those known as MCPA, 2,4-D and 2,4,5-T.

Auxins are concerned in a bewildering variety of roles in the growth of plant cells and plant organs, and in the correlation of growth in different parts of the plant. My own belief is that when the whole subject of auxins is better understood, they may prove to be as important in plant nutrition as plant nutrients themselves.

I have emphasized the effects of auxins on plant life because it is important that the reader should be aware of the value of these substances before learning that they are found in seaweed, seaweed meal and liquid seaweed extract. It is for this reason, too, that I have emphasized the value of trace elements to soil, plant and animal enzymes; for seaweed, seaweed meal and liquid seaweed extract are, in their own fields, the world's best sources of trace elements.

We have already seen that trace elements are essential to the proper functioning of enzymes and other complex organic substances, and may also have a more immediate effect on plant life—for example as unconnected molecules in plant protoplasm. We cannot set limits to the fields in which trace elements work. All we can do with any certainty is relate the presence, or absence, of trace elements to observable effects in plants and animals.

Chemical analysis of plant and animal tissue has shown them to contain almost all the known elements. We do not know, at the moment, what function is performed by gold, silver, lead, bismuth, strontium, gallium, ruthenium, and so on, all of which are found in living tissue. It may be, that as life evolved from that which was not living, it took into its tissues and mechanism the metallic components of earth and sea, so that each of these metallic components has its own essential job to do for plants, animals and men. Whether this is so or not, we do know that the concentration of trace elements essential for the satisfactory functioning of living matter is often critical. Too much of a trace element, essential in small amounts, may actually kill. Cobalt, for example, helps nitrogen fixation in plants when present at 0·006 parts per million, but restricts the growth of vegetables when present at 0·6 parts per million. Some bacilli die, or fail to reproduce, when there is a concentration of more than 50 parts per million of boron in the soil, while as little as 0·4 parts per million can cause plants to wilt. And while boron is valuable when given to sugar beet at from 10·0 to 15·0 parts per million, it can cause complete failure of a crop of oats if present in the soil at 3 parts per million. This may happen if oats are grown after a crop of sugar beet treated with a boronated fertilizer, some of which contain about 0·2 per cent of borax pentahydrate. It is now accepted that for most agricultural crops, sugar beet excepted, 3 parts per million of boron is the upper safety limit, although clover and tobacco can tolerate it up to 10·0 parts per million, and Cruciferae up to 15·0.

Trout will die if their water contains 0·1 parts per million of copper. The mould *Aspergillus niger* dies, or fails to grow, if soil contains 20 parts per million or more of manganese, while all water-culture plants die if there is 1·0 part per million of manganese in the water.

Cobalt, zinc, molybdenum and fluorine can all have toxic effects on animals if given to excess—as also can a deficiency of these elements. Examples of this are given later in the chapter on livestock. Meanwhile, tables overleaf show the effect on both plants and animals of trace-element deficiency in the case of boron, copper, cobalt, manganese, molybdenum, zinc, fluorine and iodine.

Marasmus (cobalt deficiency) in children is a failure to grow. *Azotobacter* (molybdenum) is a bacterium, *Aspergillus niger* (molybdenum) a mould. Myxoedema and cretinism (iodine deficiency) are both caused by thyroid failure.

In theory, few cultivated soils in these islands are short of the trace elements plants need. But soluble mineral salts are continually being washed out of soil by rainwater; and if the same soil is cropped year after year, this still further reduces the mineral elements present. At one time these mineral elements were replaced by natural fertilizers such as farmyard manure. Nitrogen is still replaced naturally by farmers who include a nitrogen-producing crop in their rotation, and artificially by those who give their crops nitrogen, phosphorus and potassium in the form of artificial fertilizers; but trace elements are less often replaced—and every crop removes from the soil a wide range of trace elements, some suspected of being necessary, many known to be necessary.

Trace element reserves in the soil are large, and the amount removed relatively small. It is therefore possible that for the time being, direct trace element deficiency in crops is less of a problem than indirect deficiency—the unavailability, for various reasons, of trace elements already in the soil. But it is also true that the biological cycle—whereby man returned waste animal and plant matter to the soil for further nourishment of crops whose residues were again returned, directly or indirectly, to nourish the soil—has been broken. And this biological cycle is essential for the provision of trace elements to men and animals, who depend on plants for their supply, as plants in turn depend on soil bacteria, and soil bacteria on waste matter from plants and animals. It is not possible to interfere with one aspect of this cycle, as man has done in denying plant and animal waste matter to the soil, without ultimately affecting the whole. We know what the cost of that interference has been in other directions, for example in land erosion: we have yet to find out its cost as far as the reduction of trace elements in the soil is concerned.

TABLE 10

Trace element deficiency

A table showing some of the diseases which can be caused by trace element deficiency

| *Deficient trace element* | *Soil organism* | *Plant* | *Animal* |
|---|---|---|---|
| BORON | | REPRODUCTION FAILURE<br>CORKING (apples)<br>SETTING (beans) | |
| COPPER | FAILURE OF BACILLUS TO PROLIFERATE | INADEQUATE GROWTH<br>POOR ROOT DEVELOPMENT | ANAEMIA |
| COBALT | | | PINING<br>COAST DISEASE<br>MARASMUS |
| MANGANESE | YEASTS FAIL TO GROW | FAILURE TO PRODUCE CHLOROPHYLL<br>GREY SPECK (oats)<br>YELLOW SPOT (beet)<br>MARSH SPOT (peas) | REPRODUCTION FAILURE |
| MOLYBDENUM | AZOTOBACTER<br>*Aspergillus niger* | PLANTS FAIL TO REACH MATURITY | REPRODUCTION FAILURE |
| ZINC | YEASTS FAIL TO PROLIFERATE | FAILURE TO PRODUCE LEAVES | LOSS OF HAIR<br>SOFT EGG SHELLS |
| FLUORINE | | | DENTAL CARIES |
| IODINE | | | MYXOEDEMA<br>CRETINISM |

At the moment, however, we may assume that centuries of organic farming have built up a supply of trace elements in the soil which has not yet been exhausted. A more immediate problem is the unavailability I have already mentioned—the danger that crops may be prevented from absorbing trace elements present in the soil because of other elements which come between them. This is what generally passes for trace element deficiency; and although trace element unavailability is a more correct term, it will be convenient to use the more familiar term.

In practice, where substances in the soil deny one trace element to plants, they nearly always deny others too, even if all are present in sufficient quantities. Thus soils containing lime can be manganese-deficient, although manganese is present at 100 parts per million—which is enough for any plant. If, in these conditions, there is manganese deficiency, then shortage of the trace elements of cobalt, copper, iron and zinc may also be expected. These elements, like manganese, tend to form insoluble phosphates in a limy soil; and insoluble phosphates cannot be absorbed by plants. Iron can also be 'fixed' by calcium carbonate, and thus made unavailable to plants. This is a condition sometimes encountered in orchards on limy soil.

These are natural trace element deficiencies. Similar deficiencies (as well, of course, as poisoning) result from man's actions. One example is the zinc deficiency which can result from treating soil, or plants, with mercury fungicide, and magnesium deficiency which can result from over-liming.

Attempts have been made to counter trace element unavailability by adding elements, in inorganic form, to artificial fertilizers, in particular magnesium, cobalt and manganese. Results have been inconclusive, and the general opinion in academic agricultural circles is that trace elements for plants are best supplied by animal manure in each rotation. However, there is a general shortage of these manures, and their trace element content is, in any case, dependent on that of the animals' fodder.

The effect of trace element unavailability in the soil can be overcome more effectively in two ways: by increasing the amount of organic matter in the soil; or by incorporating the inorganic trace element molecules in an organic molecule, which is then absorbed by the plant. This second method is known as 'chelating'. As to the first, it is a matter of common experience that farmyard manure, and other kinds of organic manure, help crops to grow better in conditions where there is trace element deficiency. At Haughley in Suffolk, where the Soil Association has been conducting trials, there have been no signs whatever of trace element deficiency in fields with a high limestone content—provided that humus is present. The pH in these fields (a pH figure above 7 is alkaline) is never less than 7·8, and in some cases has been as high as 9·0. In the ordinary way, alkalinity on such a scale would be bound to produce trace element deficiency by precipitating, and thus making unavailable, the manganese, cobalt, copper and iron which the soil contains. But the presence of humus has prevented this.

It is already known that micro-biological activity in the soil depends on the amount of organic matter present. It is possible, therefore, that organic manures first increase the number and activity of soil bacteria and that these bacteria then release to the plant previously unavailable trace elements.

Support is given to this theory by the results of experiments in Scotland and Russia. In Russia scientists have found that 'insoluble' compounds of calcium in the soil can be dissolved by micro-organisms. Amounts varying from 15·4 to 93 per cent of the total calcium in calcium triphosphate, from 32·4 to 49·7 per cent of total calcium in calcium phosphorite, and from 28·4 to 57·1 per cent of that in limestone, became soluble under bacterial action in trials there. Workers at the Macaulay Institute for Soil Research, Aberdeen, have also isolated an acid produced by soil bacteria which dissolves 'insoluble' calcium salts, including calcium di- and tri-phosphate. A number of minerals and soils from various places have been treated with these bacteria, and silicon, aluminium, magnesium, manganese and potassium have been released in forms in which they were available to plants.

I quoted two ways of overcoming trace element deficiencies: adding humus, and chelating. The two processes cannot really be separated. In fact, where humus encourages micro-biological activity in the soil, it is through chelating that trace elements become available to the plant; and where chelating is stimulated artificially, it is still possible that the work is done partly by micro-organisms.

A 'chelate' is an atom or molecule of metal, attached to or surrounded by organic molecules. The term is derived from the Greek word for claw. In their natural state, chelates are found wherever there is humus in the soil. In their artificial state they can be made by introducing a metallic salt to an organic material such as sugar. If an iron salt such as ferrous sulphate is used, then the iron releases its grip on the sulphur and oxygen with which it is combined in this salt, and combines with the organic molecules of the chelating material. The result is, that instead of metallic iron, or one of its salts, being offered to plants—and being refused—an organic material is offered, which includes molecules of iron. The organic material is accepted, and its iron content with it. The same process can take place with magnesium, manganese, copper, and other trace element metals.

I have made a number of references to the work of soil bacteria and micro-organisms without saying what these organisms are. They are, for one thing, infinitely small. Sir John Russell, an authority on these matters, has said that a salt spoon of soil can contain ten million of them. They do a number of things. They produce nitrogen and phosphorus in a form plants can accept. They break down the organic matter in humus—and in dead leaves and animal matter on the surface of the soil—and form soluble salts which are taken up, in solution, by the roots of plants. Like all forms of life they excrete waste products, some known as uronic acids. These improve the physical composition of the soil and, by encouraging crumb structure, make it easier to work, and better able to produce good crops. While there are some bacteria in the soil which are harmful to man—tetanus is an obvious example—those we have described are wholly beneficial both to plants and human beings.

So far we have considered plants, and in particular plants which provide food for animals and man, from a general point of view. It is now necessary to make a distinction between the growth of arable and of horticultural crops. I use the words 'arable' and 'horticultural', but it would be more accurate to think in terms of cereals (rather than of arable farming) and of vegetables (rather than of horticulture). So stated, the essential difference between the two types of crop is this: cereals can be grown profitably in a soil low in organic content; vegetables cannot. The agriculture schools tend to the belief that cereals can be grown on most soils with the addition of nothing but chemicals; and it is true that barley, for example, will produce a crop of sorts from soils with an organic content of less than 3 per cent. No vegetable crop would grow, let alone be profitable, in soils so organically starved. Further, in experiments at the National Vegetable Research Station at Wellesbourne in Warwickshire, organic manure in the form of farmyard manure has consistently produced higher yields of vegetables than mineral fertilizers alone; and with seeded crops has had a marked effect on the growth rate of the seedlings. To take only one example, carrot seedlings in soil treated with farmyard manure were found to contain more potassium than those in soil which had been given mineral fertilizers—including potassium sulphate—only. It is this greater uptake of potassium by seedlings in soil treated with farmyard manure which is held to be largely responsible for their more rapid growth.

It is possible that the differing root systems of cereals and vegetables are responsible for these reactions. If all the roots of one stem of wheat were put end to end they could easily extend to half a mile. These roots, many of them almost invisible hair roots, can push themselves out to incredible distances; and thus, delving deeply into the soil, can tap water and nutrients denied to the differing root systems of vegetables. This is why wheat can be grown successfully in the arid plains of central Canada, where there is no rain from May to August.

Vegetable roots do not go out as far as cereal roots, and remain nearer the surface of the soil. They also invest the smaller area in which they grow more densely. As a result vegetables depend, more than cereals, on the quality of the soil in which they grow; and this in turn depends, or in the past has depended, on large supplies of organic manure—the market gardens of Kent living on dung from London, those of south Lancashire on the dung from Manchester and neighbouring towns. It may therefore be said that vegetables are also conditioned to soils with a good organic content.

There is a further difference between vegetables and cereals which deserves mention. Whereas cereals are subject to a limited number of pests, most of them easily controlled, vegetables and nursery crops are subject to a host of mildews, rusts, fungi, viruses, caterpillars, cut-worms and aphides. All are more difficult, and more expensive, to combat artificially than those which attack cereals; but there is evidence that the battle against at least some of these pests is more successful if the soil has a high organic content. There is also evidence, as we shall see later, that if seaweed forms part of the organic content of the soil, or is offered to the plant in the form of a foliar spray, the fight against these various vegetable pests may be even more successful.

This brings us to the technique of foliar spraying, or giving plants additional food by spraying it on their leaves. Although a plant's ability to absorb moisture through its leaves has been guessed at, it was not until recently that its ability to absorb nutrients in the same way was suspected—although it has been known for at least fifteen years that trees can absorb nitrogen from rain, particularly in thunderstorms. Before then, it was thought that a plant's roots were the only organ which took up food, and that leaves were concerned only with photosynthesis, transpiration and, possibly, the taking up of small amounts of moisture.

Now, however, it has been proved by the use of radio isotopes that the stems, leaves and fruit of plants can absorb nutrients in solution —and that rain and mist can also leach away those nutrients through the same channels. As a result, foliar spraying has become a commercially valuable way of presenting plants, and particularly fruit trees, with easily absorbed nutrient through their leaves, thus bypassing the soil.

Nutrients in solution are absorbed through a plant's leaves more speedily than through its roots. It has been found, to take one example, that when nitrogen in the form of urea is sprayed on a large range of horticultural plants, 50 per cent of the nitrogen is absorbed within twenty-four hours. Nutrients offered in the form of a foliar spray are not only more quickly available; once absorbed, they are more effective. Eight lb. of zinc sulphate applied in a foliar spray have been found as effective as 100 lb. of the same salt applied to the soil.

Foliar spraying has other advantages. The salts of some elements essential to plant nutrition—phosphorus is one—may be reluctant to leave the soil and enter the roots of a plant. They do not show the same reluctance if offered to the same plant in the form of a foliar spray.

Foliar spraying can also act as a tonic, encouraging plants to take up more nutrients from the soil than they would otherwise have done —even, perhaps, to extract from the soil nutrients of a type not normally available to their root systems. Foliar spraying is of limited value when the plant is already well fed through its roots, or when the plant's growth is limited, for whatever reason, by outside factors. But when cold, drought, or soil deficiencies limit the amount of food which can be taken up by a plant's roots, and the plant's leaf area is large, then foliar spraying can be of vital importance in the growing of many commercial crops—from violets to walnut trees.

# 6

# SEAWEED MEAL AND LIQUID EXTRACT

In the last chapter we offered a good deal of complex information on the growth of plants. We did this because it is impossible to understand how seaweed, which has its own trace elements, its own auxins and growth-promoting substances, is of value to plants, unless we also know how plants grow—unless we know, for example, what trace elements and auxins are, or what is the value of foliar spraying.

We shall relate seaweed to these complexities of plant growth in greater detail later. In the meantime we must give a more detailed description of the two main forms in which processed seaweed can be used agriculturally and horticulturally: seaweed meal and liquid seaweed extract.

The seaweed meal and liquid extract used in agriculture and horticulture are made from brown seaweeds which grow in temperate waters—and particularly densely on the coasts of Europe north of Brittany. Three types are used: *Ascophyllum nodosum*, *Fucus vesiculosus* and *Fucus serratus*. These seaweeds grow between low and high water marks and the first two are popularly known as 'bladder wrack', although botanists usually apply this term to *Fucus vesiculosus* only. Of the three, *Fucus serratus* is the only one without air sacs.

Seaweed used for meal has its moisture content reduced by artificial drying from about 75 per cent to between 9 and 14 per cent. This is done in a conventional crop drier, either on a moving belt which travels through the drying chamber, or on a tray which is left in the drying tunnel for a period. After drying, the seaweed is ground to a meal, which varies from a fine flour for human consumption, to a coarser meal for manure or stockfeed.

Here is a typical analysis of the seaweed meal we market under the name of 'Neptune's Bounty'. Like all natural products, seaweed varies in its composition, so that the percentages given are average. The first, general, analysis, covers the whole of the material tested. The second, mineral, analysis covers only the mineral fraction of seaweed. The figures given are for percentages of pure elements, except in the case of silicon, which is expressed as silica, or silicon oxide. In addition to these mineral elements, the following are present in parts per million: manganese 33, copper 10 and zinc 48. There are also traces of sulphur, cobalt, molybdenum, boron, vanadium, nickel, chromium, tin, silver, gold and other elements. For a complete list see Table 3.

TABLE 11

General analysis of seaweed meal

| | *Percentage of total* | | *Percentage of total* |
|---|---|---|---|
| Water | 13·0 | Chlorine | 2·6 |
| Fat | 1·5 | Iodine | 0·1 |
| Protein | 7·0 | Iron | 0·08 |
| Carbohydrate | 51·0 | Magnesium | 0·7 |
| Fibre | 6·0 | Potassium | 2·6 |
| Mineral matter | 21·5 | Bromine | 0·68 |
| | | Aluminium | 0·22 |
| | | Silica | 2·0 |
| | | Calcium | 1·4 |
| | | Phosphorus | 0·2 |
| | | Sodium | 1·6 |

Liquid extracts of seaweed are made by stirring macerated seaweed in a vat containing hot water, or by alkaline hydrolysis under steam pressure. The first is the simpler method, but it extracts only about half the essential part of the seaweed—the easily soluble portions, which are largely mineral. The second, which involves processing the seaweed inside a pressure chamber, breaks down its cell structure and sugar- and starch-like substances, and thus makes it possible to extract practically all the seaweed's essential constituents. It also facilitates the extraction of all the growth-encouraging substances in seaweed—and far from diminishing them in the process, actually seems to increase them. The reason for this is not yet understood.

The technique of hydrolysis under steam pressure is the one used by my company; and where the cost of carrying water is heavy, for example in export markets, we sell this liquid extract in dehydrated form as a powder which is easily reconstituted with cold water at the point of use. One lb. of powder in 1¼ gallons of water produces the same kind of liquid extract that we sell in the United Kingdom under the name of 'Maxicrop'. At least 30 per cent of the British production of processed seaweed is now exported in this form of dehydrated liquid seaweed extract. Sales of seaweed products to the southern hemisphere are particularly valuable, since the difference in agricultural seasons between north and south allows the British manufacturer to spread production and sales over the whole year.

We also manufacture liquid seaweed extract with added organic nitrogen in the form of urea, which gives 2 lb. of nitrogen to each gallon of extract; and liquid seaweed extract with chelated iron or manganese. The nature and functions of these extracts will be explained later.

Seaweed meal and liquid extract are made from the same varieties of seaweed, and have certain qualities in common. Both provide traces of nearly every mineral element found in the earth's crust. And both, by stimulating the action of soil bacteria, help them to release to plants the phosphorus and potash present, in ample supply, in the soils of most temperate regions.

There are also differences. Meal may take months to become fully effective in the soil as a plant nutrient and soil conditioner. Extract, which can be absorbed by the plant through its leaves, as well as through its roots, may be fully effective, as plant nutrient, within hours.

Seaweed meal takes a long time to become fully effective because the carbohydrate material it contains has to be broken down by soil bacteria before it can be used by the plant. (The same applies to crop residues like straw, if they are ploughed in after harvest.) In liquid extracts, on the other hand, the cellulose-type content of the seaweed has already been broken down, and absorption can begin at once.

The fact that seaweed meal cannot become fully effective in the soil until broken down by soil bacteria is of importance for one reason: that during this period of breaking-down, the bacteria rob the soil of nitrogen in order to carry on their work. Although they return it with interest later, the amount of nitrogen immediately available for plant growth is reduced. That is why, as we shall see later, seaweed meal should never be used uncomposted in plant pots, or unless other arrangements have been made to reduce its impact on soil nitrogen there. The danger of nitrogen robbing is reduced in the open, but even here seaweed meal should be used with care near seedlings, and as a long-term, not a short-term fertilizer. Instructions concerning amounts are given later. Here we are concerned only to point out that seaweed meal, in common with other uncomposted crop residues like straw, while returning nitrogen to the soil on a larger scale after a period, does make soil nitrogen unavailable over a short period.

But apart from this characteristic of seaweed meal as opposed to liquid extract, the most important factor in deciding which of these two seaweed products to use—or whether to use a seaweed product at all—is that of expense. Seaweed meal is not cheap. In the ordinary way it is not economic to use it on crops such as cereals, which have a low value per acre. Such crops show a profit because they are grown on a large scale, and their production is highly mechanized. It is economic, however, to use seaweed meal on crops which have a high value per acre, and do not depend for their profitability on being grown extensively. In the case of cereals, for example, the increase in yield following the application of seaweed meal at £7 or £8 an acre would do little more than cover the cost of the seaweed. In the case of intensive crops like potatoes, flowers, fruit, or hops, where the profit from each acre is higher, its use would be justified.

It is still true that seaweed meal may be worth applying, even on ground which grows cereals, if there are trace element deficiencies in the soil; such deficiencies can mightily reduce the profitability of any crop. Even where no such deficiencies exist, seaweed meal will also make cereal crops less liable to lodge and (since it has a residual effect) produce a marked increase in the yield of the following crop.

But as a general rule, seaweed meal is not for use on extensive crops. Its main value is for potatoes, brassicas, sugar beet and other root crops, as well as on vegetables, flowers, fruit and herbs. For crops such as these—and where it is used, as a soil conditioner, to improve sticky clays in flower and vegetable gardens—it should be applied at rates varying from 2–5 cwt. to the acre for farm and nursery crops; and about 4 oz. to the square yard for gardens. Where garden soil is very light, sandy, or in poor heart, the amount may be increased to 6 oz. or 8 oz. a square yard.

These are the amounts we recommend for application in the United Kingdom and comparable areas. In the United States farmers and others achieve satisfactory results with about half the amount of seaweed we recommend for use in this country. Indeed, trials at Clemson College, South Carolina (described in Chapter 8), confirm that better results are obtained there with 250 lb. of meal to the acre than with 500 lb. or, in terms of ounces to the square yard, with just under 1 oz. of seaweed meal to the square yard than with just under 2 oz.

*Seaweed meal and liquid extract*

We do not know of anywhere in this country where good results can be obtained with as little seaweed as this. But the fact that one can achieve as good results in the United States with half the amounts we recommend for general use should be borne in mind by farmers there.

Liquid seaweed extract is used as a spray on foliage, or is watered on the soil. It can be applied in concentrated or diluted form, or mixed with weedkillers, fungicides, or insecticides. When used on the soil as nutrient and conditioner it is given in dilutions of from 1 in 100 to 1 in 1,000. Once in the soil, it acts as a soil conditioner in the same way that seaweed meal does. Its effect, however, is more transient.

With liquid extract, as with seaweed meal, the same results can be obtained in the United States with half the dosages we recommend for use in the United Kingdom.

Apart from the fact that seaweed meal is a better soil conditioner than liquid extract, the main difference between the two types of seaweed product is in the speed with which they produce results. In general, seaweed meal is used where consistent, long-term results are more important than immediate results. For this reason there is a tendency for it to be used on sports turf, on grass intensively cultivated, on flower gardens, and on any land cropped intensively year after year on a commercial scale. Seaweed meal begins to affect the plant through the soil after a matter of weeks or months, and thereafter on a continuous scale.

Liquid extract, on the other hand, is used to give immediate results. When applied to the leaves, extract has one valuable effect not shared by meal applied to the soil: it causes the plant to increase its uptake of soil nutrients through the roots. The mechanism which causes this effect is not understood, but has been observed in practice by many. That is why extract is used on seedlings, flowers and fruit trees, for example, to give them an immediate boost.

Apart from these uses for seaweed meal and liquid seaweed extract, which are mentioned by way of example, and are far from exhaustive, it is only practice which can decide which is the better form in which to use seaweed for agricultural or horticultural use.

We have seen that hydrolized seaweed extract can be obtained overseas in the form of a concentrated powder for mixing with water. If the reconstituted extract is used fairly quickly, no more need be done than pour the powder into the water while stirring continuously. If, however, the solution is to be stored, or re-sold, then one part of formalin by volume (or, in hot countries, two parts) should be added to 1,000 parts of solution—in practice, one teaspoonful to every gallon. The easiest way of combining the two is to pour the formalin into the container first, followed by the solution.

The addition of formalin is necessary because all liquids of vegetable origin, including seaweed extract, have a tendency to ferment after a time. The powder itself will keep indefinitely without preservative. But since it attracts water and gets damp in the same way that salt does, it should be stored in closed containers.

Hydrolized seaweed extract in powder form is exported to Australia, Bahrein, Barbados, Belgium, Bermuda, British Guiana, Canada, Ceylon, Curacao, Denmark, Falkland Islands, Finland, France, Ghana, Gibraltar, Greece, Iceland, Italy, Jamaica, Kenya, Malawi, Mauritius, New Zealand, Nicaragua, Nigeria, Rhodesia, St. Helena, Singapore, Sweden, Switzerland, Thailand, Trinidad, United States and Zambia.

The United States, New Zealand and France are our company's biggest overseas customers, although we have customers who continue to take extract in all the countries listed. In some cases consignments have gone mainly, or only, to government research stations. We assume that if the results of tests there are favourable, recommendations may be made. This has already taken place in Bahrein, where the Department of Agriculture has carried out tests with reconstituted liquid seaweed extract on flowers, and made a number of recommendations.

In other countries we have only one or two special customers and we send supplies for their private use—to the governors of certain small colonial territories, for example, or their gardeners.

In the United States, New Zealand and France, extract is used for as many purposes as in the United Kingdom, and probably for several more. In the United States, where our extract is sold as 'Sea-Born', it is consistently effective in smaller doses than are used in Europe. This may be because the Americans can count on a more stable and predictable climate: growers who spray there can be fairly confident that unexpected showers of rain will not come to wash away what they have applied. An important use there is on fruit. In the South of France, extract is used on a large scale on flowers, many of which are cut and sent by air to London and New York. Canada imports very little extract, most of it being used on forestry nurseries in Nova Scotia. Where extract is thus used in plantation work the object is to give seedlings a better start and to bring them to the planting-out stage earlier. It is seldom given to mature plants.

### *Seaweed meal and liquid extract*

Extract is used on citrus in British Guiana; on citrus and grapes in Greece; on orchids in Belgium; on flowers in Bahrein; and on market garden crops round Bangkok in Thailand. In Iceland it is used in commercial horticulture, particularly on glasshouse crops warmed geothermally. In Bermuda, Curacao, Jamaica, Rhodesia and Gibraltar its main use is in private flower and vegetable gardens. For the other countries listed I can say no more than that it is used 'generally'. This lack of certain knowledge about the applications of seaweed products overseas does not result from lack of interest on our part. Some distributors and users tell us in detail what they use seaweed extract for, and with what results. Others keep this information to themselves—as of course they are entitled to do.

A note about the treatment of plants and crops overseas appears at the end of Appendix A.

# 7

# SEAWEED AND PLANT GROWTH

Seaweed contains all major and minor plant nutrients, and all trace elements; alginic acid; vitamins; auxins; at least two gibberellins; and antibiotics.

Of the seaweed contents listed after nutrients and trace elements, the first, alginic acid, is a soil conditioner; the remainder, if the word may be forgiven in this context, are plant conditioners. All are found in fresh seaweed, dried seaweed meal and liquid seaweed extract—with the one exception of vitamins: these, while present in both fresh seaweed and dried seaweed meal, are absent from the extract.

We will deal first with alginic acid as a soil conditioner. It is a matter of common experience that seaweed, and seaweed products, improve the water-holding characteristics of soil and help the formation of crumb structure. They do this because the alginic acid in the seaweed combines with metallic radicals in the soil to form a polymer with greatly increased molecular weight, of the type known as cross-linked. One might describe the process more simply, if less accurately, by saying that the salts formed by alginic acid with soil metals swell when wet and retain moisture tenaciously, so helping the soil to form a crumb structure.

These brief notes cover two examples: one of the way in which seaweed helps to produce a crumb structure in the soil, another of the way in which it helps soil to retain moisture.

We have a market gardener customer at Sittingbourne in Kent who tells us that before he used seaweed meal, heavy rain used to run down his sloping plots and carry all his seedlings and fertilizers into the ditch. Since his introduction of seaweed, the structure of his silty, sandy soil has so improved that soil, seedlings and nutrients are no longer in danger of being washed away, even in the heaviest rain.

As to water-retaining characteristics, Miss Constance MacFarlane of the Nova Scotia Research Foundation told members of the Fourth Seaweed Symposium at Biarritz, in 1961: 'In the spring of 1956 I was greatly impressed with fields in the island of Jersey. This was not in any way a scientific experiment, but the results were most obvious. The year 1955 had been exceedingly dry. The only fields suitable for a second crop of hay were those which had been fertilized with seaweed. All the others had dried out, and had to be ploughed up for other crops.'

Research confirms this observation: two workers at the Agricultural Research Council's unit of soil metabolism (now disbanded) reported in 1947 that 0·1 of a gram of sodium alginate added to 100 grams of soil increased its water-holding power by 11 per cent. This is the first way in which seaweed and seaweed products condition the soil: by increasing its water-holding capacity, and encouraging its crumb structure. This in turn leads to better aeration and capillary action, and these stimulate the root systems of plants to further growth, and so stimulate the soil bacteria to greater activity.

As far as soil-conditioning is concerned—and that is all we are to consider for the moment—bacterial activity in the presence of seaweed has two results: first the secretion of substances which further help to condition the soil; and second, an effect on the nitrogen content of the soil. We will deal with these in turn.

The substances secreted by soil bacteria in the presence of seaweed include organic chemicals known as polyuronides. Polyuronides are chemically similar to the soil conditioner alginic acid, whose direct effect on the soil we have already noticed, and themselves have soil-stabilizing properties. This means that to the soil-conditioning agent which the soil derives from undecomposed seaweed—alginic acid—other conditioning agents are later added: the polyuronides, which result from the decomposition of seaweed.

The second effect of adding seaweed, or seaweed meal, to a soil well populated with bacteria, has already been mentioned briefly. It is a more complex matter, and requires consideration in some detail. Basically, the addition of seaweed leads to a temporary diminution of nitrogen available for crops, then a considerable augmentation of the nitrogen total.

When seaweed, or indeed any undecomposed organic matter, is put into the soil, it is attacked by bacteria which break the material down into simpler units—in a word, decompose it. To do this effectively the bacteria need nitrogen, and this they take from the first available source, the soil. This means that after seaweed has been added to the soil, there is a period during which the amount of soil nitrogen available to plants is reduced. During this period seed germination, and the feeding and growth of plants, can be inhibited to greater or lesser degree. This temporary nitrogen deficiency is brought about when any undecomposed vegetable matter is added to the soil. In the case of straw, for example, which is ploughed in after harvest, bacteria use up soil nitrogen in breaking down its cellulose, so that a 'latent' period follows. Farmers burn stubble after harvest to avoid this latent period, and the short-term loss of available nitrogen which causes it. But such stubble-burning is done at the cost of soil structure, soil fertility, and long-term supplies of nitrogen which ultimately would have been released from the decomposed straw.

It has been said by one authority that the latent period following the application of seaweed to the soil is one of fifteen weeks. But during this period, while there is a temporary shortage of available nitrogen, total nitrogen in the soil is being increased. This increase makes itself felt after the seaweed is completely broken down. Total nitrogen then becomes available to the plant, and there is a corresponding upsurge in plant growth.

It is therefore clear that while seaweed, in common with all organic matter, is beneficial to soil and plant, it has to be broken down, or decomposed, before its benefits are available. (I have already pointed out, but repeat it here, that liquid seaweed extract is not subject to this latent period. The nutrients and other substances it contains are available to the plant at once.)

This period of decomposition—or composting, as gardeners know it—usually extends over months. It can, however, be reduced by the use of dried blood and loam according to the technique invented by Mr. L. C. Chilcott, Brent Parks Manager. Only fourteen days of heating up are required before the mixture is used, and no latent period follows. This technique is described in detail on page 182.

So much for the soil conditioning effects of seaweed. Now a word about what I have called its plant-conditioning contents, beginning with vitamins.

Brown seaweeds, which are the ones used in agriculture and horticulture, not only contain vitamins common to land plants, but also vitamins which may owe their origin to bacteria which attach themselves to sea plants, in particular vitamin $B_{12}$. There is still some doubt about this—$B_{12}$ may be contained in the seaweed, although in some cases it is in associated bacteria. Vitamins known to be present in the brown seaweeds include vitamin C (ascorbic acid), which appears in as high a proportion as in lucerne. Vitamin A is not present, but its precursor, beta-carotene, is, as well as fucoxanthin, which may also be the precursor of Vitamin A. B group vitamins present are $B_1$ (thiamine), $B_2$ (riboflavin), $B_{12}$, as well as pantothenic acid, folic acid and folinic acid. Also found in brown seaweeds are vitamin E (tocopherol), vitamin K, and other growth-promoting substances. The unusual nature of the vitamin E in seaweed should be stressed. It has valuable characteristics (put technically, a complete set of isomers) found only in such seed oils as wheat germ oil.

Auxins in seaweed include indolyl-acetic acid, discovered in seaweed in 1933 for the first time. Two new auxins, as yet unidentified, but unlike any of the known indolyl-acetic acid types, were also discovered in 1958 in the *Laminaria* and *Ascophyllum* seaweeds used for processing into dried seaweed meal and liquid extract. These auxins have been found to encourage the growth of more cells—in which they differ from more familiar types of auxin which simply enlarge the cells without increasing their number. One of the auxins also stimulates growth in both stems and roots of plants, and in this differs from indolyl-acetic acid and its derivatives, which cause cells to elongate but not to divide. The balanced action of this seaweed auxin has not been found in any other auxin.

It has been proved at the Marine Laboratory at Aberdeen that indolyl-acetic acid and the other newly discovered seaweed auxins are extracted in increased quantities by the process of alkaline hydrolysis. We believe that much of the value of our hydrolized seaweed extract is due to this auxin content; but since the amount of auxin in the extract is scarcely enough to promote the increased growth which follows its use as a foliar spray, we think plants so treated are themselves stimulated to produce more vitamins and growth hormones than would otherwise be the case.

At least two gibberellins (hormones which simply encourage growth, and have not, like auxins, growth-controlling properties too) have been identified in seaweed. They behave like those gibberellins which research workers have numbered $A_3$ and $A_7$—although they may in fact be vitamins $A_1$ and $A_4$.

We now come to trace elements, some of the most important and most complex of all seaweed constituents. Two things must be said at once. The first is, that the more one studies the effect of trace elements on plants and animals, the more difficult and involved the subject becomes. Even those who devote their whole working life to the subject are far from having a complete grasp of it. The second point to make here is that while one can hope, at first, to treat trace elements separately for plants and animals, there comes a time when the two become hopelessly mixed. I shall try, in this chapter, to deal with the effect of trace elements on plants only; but some mention of their effect on animals will be inevitable, if only because animals eat plants and the trace elements they contain.

We have seen that seaweed contains all known trace elements. This is important. But it is also important that these elements are present in a form acceptable to plants. We have seen that trace elements can be made available to plants by chelating—that is, by combining the mineral atom with organic molecules. This overcomes the difficulty that many trace elements, and iron in particular, cannot be absorbed by plants and animals in their commonest forms. This is because they are thrown out of solution by the calcium carbonate in limy soils, so that fruit trees growing in these soils can suffer from a form of iron deficiency known as chlorosis. It is for this reason that plants such as rhododendrons and azaleas, which are particularly sensitive to iron deficiency, can grow only in acid soils. In these soils, iron does not combine with other elements to form insoluble salts which the plant cannot absorb, and it is therefore more freely available.

It is true that an iron salt such as iron sulphate can be dissolved in water and the solution poured on the soil, injected into an animal, or put into its feed. But iron has such a tendency to become bound up with other elements that it is not available to plants or animals when introduced in this way. If, on the other hand, iron in the form of iron oxide is dissolved in an organic compound, there will be no fusion with other chemicals in the soil, and it will be available to the plants which need it. This is the technique of chelating which makes possible the absorption of iron by living matter.

Such chelating properties are possessed by the starches, sugars and carbohydrates in seaweed and seaweed products. As a result, these constituents are in natural combination with the iron, cobalt, copper, manganese, zinc and other trace elements found naturally in seaweed. That is why these trace elements in seaweed and seaweed products do not settle out, even in alkaline soils, but remain available to plants which need them.

Hydrolized seaweed extract also 'carries' trace elements in this way, in spite of the fact that the liquid is alkaline, having a pH of nine —in the ordinary way so alkaline a solution would automatically precipitate trace elements. This precipitation does not take place in seaweed extract because the trace elements already form part of stronger, organic, associations.

With liquid extract, this ability to chelate can be taken a stage further than happens naturally with seaweed and seaweed meal. Chelation can also be used, artificially, to cause extract to carry more trace elements than are found in fresh seaweed, in seaweed meal, or in ordinary hydrolized extract.

We have ourselves exploited these chelating properties of liquid seaweed extract by manufacturing three special types, one containing added iron, one added magnesium, and one containing the three trace elements of iron, magnesium and manganese. We have also made experimental batches with copper and boron. Most metals could be chelated in this way.

It will be remembered that liquid seaweed extract differs from seaweed meal in that it can be used directly on the plant in the form of a spray. We know that the minerals in seaweed spray are absorbed through the skin of the leaf into the sap of the plant—and not only minerals, but the other plant nutrients, auxins and so on, listed earlier. Experience further suggests that plants' needs for trace elements can be satisfied at lower concentrations if those elements are offered to the leaves in the form of a spray, rather than being offered through the soil to the roots.

It is also possible that seaweed sprays stimulate metabolic processes in the leaf and so help the plant to exploit leaf-locked nutrients—for it is known that trace elements won from the soil, and delivered by the plant to the leaf tissue, can become immobilized there. And if, as has been suggested by E. I. Rabinowitch in a standard work on photosynthesis, a 'considerable proportion' of photosynthesis is carried out by bacteria at the leaf surface, spraying with seaweed extract at this point may feed and stimulate them, and thus increase the rate of photosynthesis.

We now come to the debatable matter of antibiotics in seaweed—debatable, not because there is any doubt that seaweed contains therapeutic substances, but because the precise nature of those substances is unknown. We call them antibiotics for convenience.

It is known that plants treated with seaweed products develop a resistance to pests and diseases, not only to sap-seeking insects such as red spider mite and aphides, but also to scab, mildew and fungi. Such a possibility may seem novel, but it is in keeping with the results of research in related fields. The control of plant disease by compounds which reduce or nullify the effect of a pathogen after it has entered the plant is an accepted technique. It is in this way that streptomycin given as a foliar spray combats fireblight in apples and pears, and antimycin and malonic acid combat mosaic virus in tobacco. The subject of controlling plant disease by introducing substances into the plant itself is known as chemotherapy, and is dealt with in a useful round-up article in the *Annual Review of Plant Physiology*, 1959, by A. E. Dimond and James G. Horsfall of the Connecticut Agricultural Experiment Station, New Haven, United States.

As far as chemotherapy through seaweed is concerned, the annual report for 1963 of the Institute of Seaweed Research stated that trials in which soil-borne diseases of plants were reduced by adding seaweed products to the soil were the first recorded instance of the control of disease by organic manure. 'Hitherto', the report ran, 'the majority of agricultural scientists believed that the value of organic manures was restricted to their nitrogen-phosphorus-potassium content, with perhaps some additional value as soil conditioner. This new discovery challenges this over-simplified view of the value of organic manures, and has initiated a new appraisal of this very complex problem.'

The reason why seaweed and seaweed products should exert some form of biological control over a number of common plant diseases is unknown. Soil fungi and bacteria are known to produce natural antibiotics which hold down the population of plant pathogens, and when these antibiotics are produced in sufficient quantities they enter the plant and help it to resist disease. The production of such antibiotics is increased in soil high in organic matter, and it may be that seaweed still further encourages this process.

I am aware that the claims made here, and elsewhere in this book, for the control of diseases by seaweed products, are supported more by the practical experience of growers than by the result of trials at research institutions. We have reported such trials as have taken place, but they are few in number. I cannot accept that the testimony of hard-headed farmers and horticulturists is any less reliable than that of academic researchers. But the reader might think that my attitude has been coloured by my interest, and for this reason I would say a word or two about the evidence on which these statements are based.

I have said elsewhere in this book that the evidence of the disease-controlling qualities of seaweed came to us as a complete surprise. It was those who used seaweed extract as a foliar nutrient, or seaweed meal as fertilizer, who first discovered these characteristics, and described them to us. We make no other claims than these, only record what users say, and it would be a poor service to truth to censor this evidence of the value of seaweed because it has not been confirmed in all respects by trials at research stations. Where these trials have taken place they are later reported. Trials in this country have been few, for a variety of reasons which need not concern us. We might regret that state-supported stations noted for a high standard of scientific integrity are also conservative in outlook, and little disposed to test that which is unusual. It is not for us to criticize their choice of subjects for research, but our own information is so striking that we should wholeheartedly welcome testing of seaweed and seaweed products by those with complete facilities for doing so. The evidence we have collected would then be respectably 'scientific'—and we do not doubt that the findings would corroborate our claims to the full.

# 8

# YIELD, CROP QUALITY AND HEALTH

Evidence of the beneficial effect of seaweed and seaweed products on agricultural and horticultural products comes from this country and overseas.

*Yield*

Some of the most important evidence of increased yield comes from the United States, from trials in New Hampshire, South Carolina and New Jersey. The trials in New Hampshire, reported in the 1944 *Journal of the American Society for Horticultural Science*, concerned McIntosh apple trees in an orchard at Durham. Each tree was given one of four mulches: hay, sawdust, seaweed and grass. As the following table shows, those mulched with hay or seaweed produced more, and larger, fruit than those mulched with sawdust or grass.

TABLE 12

Apple yields and seaweed

McIntosh Red apple trials at Durham, New Hampshire, United States

| *Mulches* | *Yield of each tree in bushels* | *Average weight of fruit in pounds* | *Red colour—percentage of surface covered* |
|---|---|---|---|
| Grass | 0·27 | 0·305 | 83·8 |
| Hay | 1·16 | 0·362 | 68·4 |
| Sawdust | 0·59 | 0·311 | 85·7 |
| Seaweed | 1·24 | 0·370 | 63·6 |

The South Carolina trials took place at Clemson College of Agriculture, and covered a wide range of crops. They are worth considering in some detail.

## *Yield, crop quality and health*

Nine crops were tested in one of the main trials: sweet pepper, tomato, sweet corn, peas, sesame, cotton, soya bean, lima bean and okra (a plant cultivated for its fruit pods, which are used as a vegetable; sometimes known as 'ladies' fingers'). They were planted in 30-ft. square plots using five replications—each treatment, that is to say, was given to five plots in different parts of the field where the trials took place. All plots were given 1,000 lb. an acre of artificial fertilizer, in addition to seaweed meal given at two rates before planting: 250 lb. to the acre, and 500 lb. to the acre.

In these, as in all the Clemson College trials, the seaweed meal and liquid extract were made from *Ascophyllum nodosum.* The liquid extract was standard 'Maxicrop' supplied by my company, the meal was supplied by Algea Producter of Kristiansand, Norway.

This table gives percentage gains and losses from the two treatments, to the nearest whole figure.

TABLE 13

Nine crops trial at Clemson

| | *Percentage gain or loss with 250 lb. seaweed meal to the acre* | *Percentage gain or loss with 500 lb. seaweed meal to the acre* |
|---|---|---|
| Sweet pepper | 31 gain | 13 gain |
| Tomato | 37 gain | 11 gain |
| Sweet corn | 104 gain | 75 gain |
| Okra | 3 loss | 7 gain |
| Peas | 6 loss | 4 gain |
| Sesame | 17 loss | 4 gain |
| Cotton | 8 loss | 78 loss |
| Soya bean | 23 gain | 20 gain |
| Lima bean | 12 gain | no change |

Residual fertility from the previous year's application of seaweed was checked the following year for sweet pepper, sweet corn, okra, peas and lima beans. The following percentage gains or losses compared with the control plots were then recorded. The table is not entirely satisfactory, largely because of the unexplained loss of 10 per cent in the case of lima beans given 250 lb. of seaweed meal to the acre in the previous year. But the table as a whole does show, as general experience confirms, that seaweed meal is more effective at 250 lb. to the acre than at 500 lb.

TABLE 14

Residual fertility trials at Clemson

| | *Percentage gain or loss with 250 lb. seaweed meal to the acre in previous year* | *Percentage gain or loss with 500 lb. seaweed meal to the acre in previous year* |
|---|---|---|
| Sweet pepper | 19 gain | 20 loss |
| Sweet corn | 9 gain | no change |
| Okra | 10 gain | 4 gain |
| Peas | 20 gain | 24 gain |
| Lima beans | 10 loss | 18 loss |

It will be seen that in the first year of the trials, average gain for sweet peppers, tomatoes, sweet corn, soya bean and lima bean was 41 per cent at the 250-lb. rate, and 24 per cent at the 500-lb. rate; that average loss for okra, peas, sesame and cotton was 9 per cent at the 250-lb. rate, but that this loss turned into a gain at the 500-lb. rate, in all cases save that of cotton, where the loss was disastrously increased. I have discussed such losses as took place with the staff at Clemson, but have been unable to explain them. Clearly there is some very potent factor to account for so emphatic a decline, and in only one instance.

I suspect that in these trials seaweed was applied too soon before planting out, so there was not enough time for it to break down. The interpretation of the figures is also made more difficult by the fact that 1,000 lb. of chemical fertilizer, of unknown content, were applied to each acre.

The table on page 98 shows overall yield in these trials in pounds an acre.

In tomato trials at Clemson, using seaweed meal and seaweed extract, a yield increase of 61 per cent was obtained by weekly watering with a 1 in 25 solution of seaweed extract. This increase was reduced to one of 37 per cent when a more dilute solution of 1 in 50 was used. Four plants, each in a single pot, were used as controls. The trial was a small one, involving only twenty-four tomato plants, each in a single pot, and so was statistically barely adequate. The following treatments were tested: seaweed meal at 250 lb. to the acre and 500 lb. to the acre; liquid seaweed extract at rates of 1 in 25 and 1 in 50. Each of six different types of treatment was thus given to a group of four plants. The essential figures are set out on page 99.

TABLE 15

Yield trials at Clemson

Overall yield in lb. an acre after seaweed meal given at 250 or 500 lb. an acre. All plots, including control, also given 1,000 lb. of artificial manure an acre

| | *Year* | *Sweet pepper* | *Tomato* | *Sweet corn* | *Okra* | *Peas* | *Sesame* | *Cotton* | *Soya beans* | *Lima beans* |
|---|---|---|---|---|---|---|---|---|---|---|
| Control | 1960 | 8,620 | 14,519 | 1,341 | 9,269 | 5,164 | 600 | 353 | 1,200 | 3,262 |
| 250 lb. | 1960 | 11,283 | 19,922 | 2,736 | 9,035 | 4,847 | 498 | 326 | 1,470 | 3,760 |
| 500 lb. | 1960 | 9,734 | 16,144 | 2,347 | 9,943 | 5,391 | 626 | 76 | 1,440 | 3,262 |
| Control | 1961 | 3,323 | | 5,811 | 2,597 | 3,142 | | | | 3,614 |
| 250 lb.* | 1961 | 3,977 | | 6,356 | 2,869 | 3,777 | | | | 3,251 |
| 500 lb.* | 1961 | 2,651 | | 5,811 | 2,706 | 3,904 | | | | 2,960 |
| 250 lb. | 1961 | 3,686 | | 6,265 | 2,815 | 3,741 | | | | 2,740 |
| 500 lb. | 1961 | 3,450 | | 6,447 | 3,051 | 3,705 | | | | 2,143 |

* Applied in 1960.

TABLE 16

Tomato trials at Clemson

| | *Yield in grams* |
|---|---|
| Control | 2,169 |
| Seaweed meal 250 lb. to acre | 2,339 |
| Seaweed meal 500 lb. to acre | 2,136 |
| Liquid extract A | |
| 1 in 25 solution weekly | 2,768 |
| 1 in 50 solution weekly | 2,269 |
| Liquid extract B | |
| 1 in 25 solution weekly | 3,488 |
| 1 in 50 solution weekly | 2,964 |

Field trials have taken place in Norway to assess the value of seaweed as a supplementary fertilizer—that is, to see what effect it had when given to a crop already receiving artificial fertilizer containing nitrogen, phosphorus and potassium. The trials were with turnips, and two supplementary treatments were given. In one, seaweed only was added to the artificial fertilizer; in the other, the same amounts of nitrogen, phosphorus and potassium as the seaweed contained were added in place of the seaweed.

The soil, at Selbu, was sandy, newly cultivated, and deficient in phosphorus. Three crops were raised, in 1958, 1959 and 1960, with results set out below. Since the figures are comparative, they serve equally well as factors, and have therefore not been translated from kilograms of dry matter a decare. It should, however, be remembered that turnips would not usually be grown as a commercial crop in a soil of this character.

TABLE 17

Field trials at Selbu

| | *Dry matter yield of turnips in kilograms a decare* | | |
|---|---|---|---|
| | *1958* | *1959* | *1960* |
| Artificial fertilizer alone | 236 | 503 | 599 |
| 125 kilograms of seaweed meal a decare added | 365 | 677 | 605 |
| Nitrogen-potassium-phosphorus content of 125 kilograms of seaweed added | 321 | 591 | 597 |
| 250 kilograms of seaweed meal a decare added | 352 | 679 | 633 |
| Nitrogen-potassium-phosphorus content of 250 kilograms of seaweed added | 337 | 764 | 594 |

It will be seen that the addition of seaweed produced a much greater yield than basic artificial fertilizing in 1958 and 1959, although in 1960 the increase was reduced to only a few per cent. It will also be seen that double the amount of added seaweed produced slightly less than the single seaweed treatment in 1958, much the same in 1959, and more in 1960.

Where nitrogen-phosphorus-potassium equivalents are concerned, there was virtually no difference between single and double additions except in 1959, when double treatment gave the greatest yield of all.

Sverre Myklestad of the Norwegian Seaweed Research Institute, who conducted the trials, commented: 'One can assume that about 125 kilograms of seaweed meal a deçare (roughly 1,100 lb. an acre), applied to a soil of this type as supplement to good basic fertilizing, will give an increase in the crop yield over that expected from the amounts of nitrogen, phosphorus and potassium applied; and it is likely that the effect will be greatest on newly-cultivated ground.'

In a second trial with fodder beet at Sola, Norway, the results obtained with and without added seaweed varied so little that it is difficult to think that the results have significance. In a third trial, with cauliflower in frames, yield was little affected by the addition of seaweed, but more of the best quality curds were produced.

Commenting on all three trials, Myklestad says: 'The results of these experiments show that the seaweed fertilizer used has a positive effect on special soils, in addition to the expected effect from the content of nitrogen, potassium and phosphorus. On soil in very good condition the effect of seaweed fertilizing was found to be small when given as supplement to a good basic fertilizing.'

Certain comments may be made on these trials, and were made when they were reported to the Fourth International Seaweed Symposium at Biarritz in 1961. The first is that seaweed, which is a water-retaining substance, may have affected yields as a result of rainfall at certain periods; the second is that using seaweed at such a low rate can never make a poor soil fertile.

Nearly all the evidence for increased crop yields in this country following the use of seaweed products comes, as I have said, from farmers and horticulturists. One large field trial did take place in Bedfordshire in 1965 with the help of the National Agricultural Advisory Service; it showed a dramatic increase in brussels sprouts yield, after use of seaweed extract, and is worth describing in detail.

The trial was interesting too because it did not set out primarily to assess the effect of seaweed extract on yield, but to see whether there was any truth in repeated claims that small amounts of seaweed extract given as a foliar spray reduced infestation by the grey cabbage aphis. As it turned out, there were few aphides on the crop in 1965, and the counts made were of little significance. But yields of marketable sprouts were weighed separately for each treatment immediately after picking, with the following result. With demeton-methyl sprays alone there was a 6·2 per cent yield increase over untreated controls; with demeton-methyl and 1 in 20 extract applied 'when demeton-methyl was required', there was an increase of 11·0 per cent; with 1 in 120 extract alone at three-week intervals, an increase of 12·4 per cent; and with demeton-methyl and 1 in 120 extract at three-week intervals, an increase of 33·0 per cent. The pickers also declared unanimously that quality and colour of sprouts from seaweed-treated rows were noticeably better than those from other rows.

Demeton-methyl, brand name Metasystox, was given as a dilution of 16 fl. oz. in from 60–70 gallons of water. The seaweed extract 'Maxicrop' was given, both alone and in combination, at the rate of 2 gallons to the acre.

The trial took place at Sandy. The trial area, some 3¾ acres in extent, was a uniform section forming part of a large field of sprouts. There were 150 rows in the section, each some 120 yards in length. These were first divided into three equal blocks of fifty rows, and then further divided into five strips ten rows in width. Each block was given five treatments, one treatment for each of the five strips.

The time of application of treatments 3 and 5, shown in Table 18 as 'given when necessary', was left to the spraying contractor. In both cases, applications took place on 28th July and 14th August. Spraying for the three-week treatments (2 and 4 in the table) took place on 23rd July, 14th August and 3rd September. The late start and irregularity of spraying are accounted for by the wet summer. After the fourth pick had been taken, the crop was ploughed in.

On the basis of the figures tabulated overleaf, increased yield in a 5-ton/acre crop would be about 12 cwt. The three applications of seaweed extract cost about £3 an acre, so that to produce an extra yield of 12 cwt. for this price is a profitable transaction. The same remarks apply with greater force when a combination of seaweed and demeton-methyl gives even higher yields—and in this case, when extract is added to insecticidal spray, no additional cost (save that of materials) is incurred.

The table shows the effect of 'Maxicrop' seaweed extract and Metasystox demeton-methyl, alone and in combination, on the sprout crop. Yield is given in pounds a strip.

TABLE 18

Seaweed and brussels sprouts

Yields in lb. a strip in Bedfordshire trials

| | *1* | *2* | *3* | *4* | *5* |
|---|---|---|---|---|---|
| | *Control* | *'Maxicrop' 1-in-120 at 3-weekly intervals* | *Metasystox only, applied when necessary* | *Metasystox plus 'Maxicrop' 1-in-120 at 3-weekly intervals* | *Metasystox plus 'Maxicrop' 1-in-120 applied when Metasystox required* |
| | 1,943<br>1,955<br>1,156 | 2,000<br>2,249<br>1,430 | 2,225<br>2,124<br>1,018 | 2,480<br>2,425<br>1,796 | 2,430<br>2,013<br>1,166 |
| Total yield | 5,054 | 5,679 | 5,367 | 6,701 | 5,609 |
| Percentage increase | — | 12·36 | 6·19 | 32·98 | 10·98 |

Considerable increases in the yield of potatoes have resulted from the use of extract on farms in Lincolnshire and Cambridgeshire. On deep silt in Lincolnshire an average increase of 6·8 per cent with standard extract, and one of 18·9 per cent with extract and chelated iron, have been recorded. On black fen in Cambridgeshire the figures were, with standard extract, a 20·2 per cent increase, with extract and chelated iron, a 13·5 per cent increase. Three adjoining strips, each six rows in width, and each about half an acre in extent, were used for the trial on each site. All strips were given artificial fertilizer at the rate of 10 cwt. to the acre. The composition of this fertilizer, although it differed between the two sites, was the same for all strips on one site. One of the three strips on each site was left untreated as control. The second was given one application of standard seaweed extract at the rate of 10 gallons an acre. The third was given one application of extract and iron at the same rate. Guard zones between the rows prevented any overlap as a result of spray drift.

Yields of square-yard plots at fixed intervals throughout the whole strip were weighed at lifting. They gave control averages of 6·7 lb. of potatoes to the square yard for Lincolnshire silt, and 8·4 lb. for black fen. The heavier yields from treated strips were then used, with the appropriate control figure, to provide the percentage increases earlier listed.

On the basis of these figures it can be assumed that the use of 'Maxicrop' standard or iron-chelated extract will give increases in potato yield of from 7 to 20 per cent an acre.

It will be noticed that extract with iron gave higher yields than standard extract on silt. This may have been because of increased resistance to drought. On the other hand, it may be that the result on organic soil—black fen is dried-out peat—is the unusual one, rather than that on inorganic silt soil. Undrained fen peat often contains deposits of what are called bog iron, which remain in the soil when it is drained. Perhaps the addition of chelated iron to a soil already containing traces of this 'organic' iron is less effective than its addition to an inorganic silt soil containing no 'organic' iron. This is, of course, yet one more indication of the necessity for extensive tests by a totally disinterested laboratory team; I should like to see trials made, on soils known to contain bog iron, with other chelated minerals involved.

Increased tomato yields have also been recorded following the use of a special liquid seaweed extract which contains added nitrogen, phosphorus and potassium. When this extract, prepared specially for tomato growers, was fed to 700 tomato plants on a nursery at West Winch, Kings Lynn, they produced 4,698 lb. of fruit, an average of 6·7 lb. a plant. Nine hundred plants on the same nursery, fed potassium nitrate and urea, produced 4,422 lb., an average of 4·9 lb. a plant. The fortified seaweed extract was fed from 20th June to the middle of July in a 1 in 1,000 concentration, and then until the second week in August in a concentration of 1 in 750.

Among a small number of trials testing seaweed and seaweed products at government experimental stations in this country have been those at Long Ashton Agricultural and Horticultural Station, Bristol, and those at Rosewarne Experimental Horticulture Station, Camborne, Cornwall.

In the first of two experiments at Long Ashton, ten Baldwin blackcurrant bushes sprayed with seaweed extract gave an average yield of 382·8 lb., compared with one of 339·9 lb. for ten untreated control bushes. In the second, average figures were 176·6 lb. for twelve treated bushes and 145·6 lb. for twelve untreated bushes. The interesting thing about these increases—of 12 per cent in the first trial and 27 per cent in the second—is that they were obtained with so little extract. Three applications of 1 in 400 solution were given at fortnightly intervals, beginning at the first open-flower stage—and the total amount of undiluted extract given to each acre was only 1½ gallons.

In trials at Rosewarne to compare the value of farmyard manure, seaweed compost, straw compost, straw, and no organic manure on winter cauliflower between 1955 and 1962, seaweed compost gave an average, over the eight years, of 281 marketable heads an acre; farmyard manure, 276; straw compost and straw, 273 each; and no organic manure, 262. I agree with the observation which most readers are already making—the differences, even between the extreme numbers 281 and 262, are not large enough to justify extravagant claims. Small local differences could account for minor variations, so could degrees of skill in handling at all stages; nevertheless, the seaweed compost produced the best figure out of five and with an advantage of a score over the 'no organic manure'.

In trials at the same station, farmyard manure, and seaweed composted with straw gave the best crops of early potatoes and leeks.

Such trials, as I have already written, have been all too few. Practical experience, though, has proved that all crops (and particularly root crops) respond well to seaweed. To give one example from each of three main crops, I can say that sugar beet at 20 tons to the acre with a 15·1 per cent sugar content has been raised by a Wainfleet St. Mary, Lincolnshire, farmer on 16 acres of land treated with 5 cwt. of seaweed meal to the acre. On 145 acres not so treated, yield was 18·8 tons an acre, with a 14·1 per cent sugar content. The average value of the treated crop was £120 an acre, of the untreated, £105. In addition, all the beet *excepting* that given seaweed meal had to be sprayed twice against 'virus yellows' at a cost of 70s. an acre.

Almost 100 per cent increase in the yield of carrots has been obtained by a Marham, Norfolk, farmer after spraying with seaweed extract at 6 gallons to the acre. The part of his 50-acre field which was sprayed produced a crop estimated at 25 tons to the acre—12 tons to the acre more than the rest of the field. This is one of the most spectacular results so far recorded.

Potato yields on a farm at Outwell in the Isle of Ely also increased 27 per cent after being sprayed with extract at 4 gallons to the acre on 30th June and 14th July 1959.

We have received particularly good yield reports in the case of runner beans after the use of seaweed fertilizers—among them increases of 3 tons an acre with 'Streamline', and of 1½ tons an acre with 'Blue Lakes' dwarf runner beans. These, and the reports of increased root crops, are chosen from hundreds of verbal reports from our representatives. Individual reports from growers may be suspect, but cumulative evidence from different sources spread over ten years can with fairness be accepted. And while the use of seaweed fertilizer on extensive crops is not really economic unless there is (for example) serious trace element deficiency, I may mention one field experiment in Suffolk where wheat sprayed with extract, to which nitrogen had been added in the form of urea, produced 2¼ cwt. an acre more than the area of wheat not sprayed. Two fields, each of 20 acres, were involved, both having the same crop, yield and cultivation in the previous year. The farmer gave 3 cwt. an acre of a proprietary chemical fertilizer to both fields. In addition he gave to one field a 2½ cwt. dressing of nitrochalk, to the other a 1¼ cwt. dressing only; but it was this second field, which produced the heavier yield of wheat, which also received 2 gallons an acre of seaweed extract plus nitrogen.

Seaweed also plays its part in competition growing. John Cooke of Funtington in Sussex, an enthusiastic seaweed user for many years, and for four years running winner of the *Garden News* prize for the heaviest crop of potatoes grown from six sets, used seaweed products —extract until 1964, thereafter seaweed meal as well—to achieve his outstanding results. In 1961 he grew 206 lb. from six sets; in 1962 (when the haulms were 9 ft. high) 351 lb. 2 oz.; in 1963, 391 lb. 3 oz.; in 1964, 500 lb. 2 oz.; and in 1965, 563 lb. Having achieved this result, which he believed to be a world record, Mr. Cooke announced that he was retiring from the championship potato business. He used bone meal, wood ash, straw and compound fertilizer as well as cow, poultry and pig manure with his seaweed treatments. These last included soaking the seed in a 25 per cent solution of extract for two hours once a fortnight from October to March, when planting took place; adding seaweed meal to the plot; and watering soil and spraying haulms weekly with liquid seaweed extract.

The improvement which takes place in flowers following the use of seaweed fertilizer is a matter of personal observation rather than objective measurement. But trials with poinsettias at Clemson College, South Carolina, over two seasons showed that the number, size, weight and quality of the flowers were improved by seaweed extract. The flowers, 'Barbara Ecke Supreme', were given extract in concentrations of 1 in 5, 25 and 50, and meal at 300, 600 and 900 lb. an acre. Only when extract was given at a concentration of 1 in 5 was no improvement recorded. This was too strong for the flowers, was said to 'stun' them, and was discontinued after the first application. The size of flowers was also reduced by treating them with extract at a concentration of 1 in 25. But in all cases save that of the 1 in 5 solution, flowers had a higher dry matter content. It was also found that when the poinsettias were laid under the benches from December to May in readiness for new growth, those treated with seaweed had thicker and more stocky cuttings, which rooted more easily and therefore produced better plants for the Christmas trade.

Eleven poinsettia plants were used for each of seven treatments. The trial began on 26th October, and extract was applied weekly until 12th December. Meal was applied once at the beginning of the trial, when it was worked into the top inch of soil. This table gives detailed results of the trials.

TABLE 19

Poinsettia trials at Clemson

| | *Average number of flowers on each plant* | *Diameter of flowers in inches* | *Average weight of flowers in grams first season* | *Average weight of flowers in grams second season* | *Percentage dry matter content* |
|---|---|---|---|---|---|
| Control | 3·8 | 14 | 92 | 138 | 13·7 |
| Seaweed extract 1 in 50 solution | 4·7 | 14 | 106 | 166 | 14·2 |
| Extract 1 in 25 solution | 4·5 | 13 | 92 | 140 | 13·9 |
| Extract 1 in 5 solution | 4·4 | 12 | 81 | 114 | 12·8 |
| Seaweed meal at 300 lb. an acre | 4·5 | 14 | 123 | 189 | 14·0 |
| Meal at 600 lb. an acre | 4·9 | 14 | 126 | 194 | 14·0 |
| Meal at 900 lb. an acre | 4·5 | 14 | 119 | 180 | 13·7 |

*Germination*

Liquid seaweed extract encourages germination, often in an outstanding way, as tests at Clemson College have shown. Eighty-four per cent of beet seed soaked in extract there germinated after a week, of untreated seeds, none. It was also found that soaking seed for thirty minutes at, or just before, germination, increased germination rate by a further 25 per cent.

The respiratory activity of a number of different types of seed was also greatly accelerated as a result of soaking them in varying concentrations of seaweed extract. Seeds treated were zinnia, tobacco, pea, turnip, tomato, radish, cotton, white pine, loblolly pine (a kind of pitch pine found in the south-east of the United States), privet, heavenly bamboo (an evergreen flowering shrub) and American holly. In all cases, and at all concentrations—which varied from pure extract to a solution of 1 in 500—the respiratory activity of the seed was increased. The higher the concentration, the higher the respiratory activity, although there was then a corresponding fall in the numbers of seed which germinated. Concentrations which gave a moderate increase in respiratory activity did, however, give increased germination too. The point at which balance was achieved varied with the type of seed, but it was usually with solutions of 1 in 25 and 1 in 50. It was also found that germinating tomato seed soaked in extract produced plants which grew better, and were of healthier colour. Pepper seed so treated produced plants which set fruit at an earlier age, and gave a better yield.

These results are confirmed by the findings of laboratory tests with grass seed in the United States. In this case 'Sea Magic 3' seaweed extract (Chase Organics Limited, Shepperton, Middlesex) was used to germinate the creeping red fescue used for seeding road verges. The trials were carried out by two members of the staff of the Connecticut State Highway Department Laboratory, who reported a dramatic increase in the speed of seedling emergence. The seed was placed on sand. In order to simulate field conditions, some was left uncovered, some was covered lightly with sand, and some covered with too much sand. Equal numbers of seed were left untreated, or treated with varying strengths of seaweed solution.

A dramatic increase in speed of emergence was obtained with seed treated with 0·5 and 1·0 per cent solutions of extract, and for one week the rate of growth of these seedlings was faster than that of the control. After twenty-five days, seedlings given 1·0 per cent solution began to show the effect of plant food deficiency more than those given 0·5 per cent solution, and from the twenty-fifth day began to die off rapidly. The following table of emerged and living seedlings shows the speed of emergence with the two strengths of solution, and the deterioration which took place after twenty-five days at the greater concentration.

TABLE 20

Germination trials, Connecticut

Effect of 'Sea Magic 3' seaweed extract on germination of creeping red fescue

| Days | 7 | 8 | 11 | 14 | .. | 25 | 28 | 30 | 35 |
|---|---|---|---|---|---|---|---|---|---|
| Number of seedlings living in control conditions | 11 | 14 | 22 | 29 | .. | 45 | 48 | 49 | 55 |
| Number of seedlings living after treatment with 0·5 per cent solution | 18 | 24 | 37 | 41 | .. | 57 | 61 | 60 | 65 |
| Number of seedlings living after treatment with 1·0 per cent solution | 27 | 36 | 48 | 52 | .. | 62 | 58 | 52 | 29 |

Seed treated with solutions of seaweed extract stronger than 5·0 per cent in these Connecticut trials had their growth retarded, while at an 18 per cent concentration, the highest tested, no seedlings emerged, and the seeds were dead after twenty-one days. Good emergence results were not obtained if seedlings soaked in solution were dried before planting.

Growth trials with 'Sea Magic 3' have also taken place in this country, using mustard seed sown in vermiculite. These were conducted by S. B. Challen and J. C. Hemingway of the Portsmouth College of Technology, and are described in the *Proceedings* of the Fifth International Seaweed Symposium.

The seeds were given either distilled water; 'Sea Magic 3'; 'Sea Magic 3' with added nitrogen, phosphorus, potassium and magnesium; or a synthetic fertilizer containing these elements. 'Sea Magic' was given in a 1 per cent solution in each case, and the synthetic fertilizer was diluted so that it gave the same concentrations of nitrogen (0·06 per cent), phosphorus (0·046), potassium (0·035) and magnesium (0·003) as that in the seaweed extract with these chemicals added. (There was no attempt, as in the Connecticut trials, and in others later described, to assess the retarding effects of a stronger seaweed solution.)

The greatest growth (to 5·91 cm. in height) was recorded with 'Sea Magic' and added chemicals, and the next highest (4·84) with 'Sea Magic' alone. Synthetic fertilizer produced a growth of 3·84 cm., and distilled water alone one of 3·49. This is the most significant finding of those described in detail in the *Proceedings* of the Seaweed Symposium.

Comparable results with flowers and vegetables were obtained by Dr. J. A. Mowat in trials in Aberdeenshire in 1961 and 1962. The trials took place in February and March, in a propagator with soil warmed to 60° F. Seeds were sown in boxes containing John Innes No. 1 compost, and watered with solution weekly. It was then found that the germination and growth of cabbage, brussels sprouts, leek and lupin seed were retarded by a 10 per cent solution of liquid seaweed extract; the germination of mustard, cress, annual dwarf phlox and brussels sprouts, was improved by a 1·0 per cent solution. (John Innes No. 1 compost is made of seven parts by loose bulk of sterilized loam, three of horticultural peat and two of coarse sand. To every bushel is added 4 oz. of John Innes base made up as follows: 2 parts by weight hoof and horn, 2 superphosphate, 1 sulphate of potash and ¾ oz. ground limestone or chalk. John Innes No. 2 is the same, but with twice the base and chalk and John Innes No. 3, the same, but with three times the base and chalk.)

Trials with a number of varieties of late chrysanthemum cuttings also took place in January and February 1961 and 1962. These were rooted in John Innes No. 1 compost in a propagator, and heated to between 50° and 60° F. Treated cuttings were watered weekly with a 1 per cent solution of seaweed extract, controls with water only. After two or three weeks most of the treated plants were a brighter green, and showed more vigorous transpiration. In the case of one variety, 'Autumn Tints', both treated and control plants were well rooted. Treated 'Strawberry Roan' cuttings were vigorously rooted, although the roots of untreated cuttings of the same variety were only just beginning to develop. Treated cuttings of the variety 'J. R. Johnston' had much longer roots than the control, and there was also an obvious difference between treated and untreated varieties of 'Doris Rams', 'Susan Ayles' and 'Elizabeth Edney', with the treated plants in far better shape.

When the cuttings were well rooted, they were put into peat pots containing John Innes compost No. 2, later into clay pots with compost No. 3, and planted out at the beginning of June. While in pots, treated plants were watered weekly with 1 per cent solution, and after being planted out, were sprayed once a fortnight with 1 per cent solution. Treated plants, returned to the greenhouse in October, were taller than the untreated, and flowered a week earlier.

Perhaps one of the most dramatic examples of the value of seaweed as a rooting agent came when a 0·5 per cent solution of seaweed extract was used in comparison with a commercial hormone preparation recommended for rooting seedlings. This was in trials with honeysuckle cuttings conducted by Dr. J. A. Mowat. The trials began on 14th February 1962, a soil-warmed propagator frame being used. Of the twenty-five cuttings treated with extract, fourteen took root, but of the twenty-five treated with hormone preparation, only one. It was also noticed that leaf buds, which opened on the seaweed-treated cuttings, failed to open on those treated with the commercial hormone preparation.

Trials with other woody cuttings such as berberis, beech and lilac, were inconclusive.

### *Trace element deficiency*

Trials with tomato plants in New Zealand have confirmed that seaweed not only provides trace elements, but makes it possible for the plants to take up more of these elements from the soil. The tomatoes in these trials were treated with seaweed containing between four and six parts per million of manganese. When the leaves of these treated plants were analysed, they not only contained more manganese than those of untreated plants, but they also contained more manganese than was present in the seaweed. This shows that the seaweed made soil manganese available to the treated plants.

TABLE 21

Trace elements and tomatoes

Manganese content in parts per million of tomato plant leaves grown with and without 1·25 per cent added seaweed meal in New Zealand trials

| *Soil* | *No seaweed* | *Seaweed* |
|---|---|---|
| Sand | 343 | 315 |
| Clay A | 100 | 214 |
| | 138 | 343 |
| | 138 | 489 |
| Clay B | 1,588 | 2,642 |
| | 900 | 1,792 |
| | 900 | 1,706 |
| | 300 | 1,440 |
| | 160 | 1,050 |
| Potting soil | 104 | 181 |
| Peat | 248 | 1,488 |

The tomatoes were grown in sand, two types of clay, potting soil and peat, to which 1·25 per cent of seaweed meal had been added. The table on page 110 shows the manganese content of leaves in parts per million from treated and untreated plants.

Comparable results have been obtained in the United States. In 1955 an American research worker gave poorly bearing lime trees, suffering from deficiencies of iron, zinc, magnesium and boron, 200 lb. of fresh seaweed each. Two years later he found the trees were bearing profusely. They also showed a healthy green colour, which suggested that their trace element deficiencies were at an end. These deficiencies were not made up by the seaweed only. Trace elements in the soil, previously unavailable to the trees, were released by the seaweed and by the microbiological activity it encouraged.

Seaweed extract containing chelated minerals produces similar results. As a result of using extract containing 1 per cent chelated iron we have been able to grow camellias, dwarf rhododendrons, *Lilium longiflorum*, *Lilium auratum*, and a number of varieties of *Azalea indica* in our garden at Holdenby, where the soil has a pH of eight. In the ordinary way a soil so alkaline would not grow these plants. At the time of writing, however, they have been growing well for three seasons, producing slightly better blossoms than controls grown in peat compost with ordinary seaweed extract. I describe our experiences with these plants in greater detail on page 191.

I also know of a number of pear orchards on soils with a pH of eight which have responded well to heavy doses of extract with chelated iron. One in particular, on alkaline soil in Norfolk, had a long history of chlorosis; but after being treated in the spring with extract and chelated iron at the rate of 13 gallons to the acre, showed an increase of 500 per cent in leaf iron by mid-August.

### *Soil bacteria*

We have described the action of seaweed, both in meal and liquid form, as catalytic; and in the previous section on trace elements, observed that seaweed helps plants to take up nutrients from the soil without itself providing nutrients on anything like the same scale.

This increase in the availability of nutrients in soils treated with seaweed products, or in soils supporting seaweed-sprayed plants, may result from a number of causes. But whether auxins, or trace elements, or both, are responsible, it does seem as if their effect is indirect. It would appear that these constituents of seaweed have a stimulating effect on plant growth and yield because they first stimulate the growth and activity of soil, or plant, bacteria.

We tend to ignore the fact that we, and all other animals, depend for our life on plants; that plants depend on soil; that soil, to be useful and fertile, depends on bacteria; and that the whole of this interdependence makes up the biological cycle to which we have already referred. We do not know much about soil bacteria, but that does not prevent our interfering with them on a massive scale. We not only ignore the need to feed them by failing to return organic residues to the soil; we actually destroy them at times by chemical cultivating, and similar fashionable agricultural techniques.

Do seaweed products, when they are used, produce dramatic results partly because they minister to the needs of these bacteria? For seaweed products, far from poisoning soil bacteria, help to feed them. Seaweed can even support its own bacteria before being added to the soil. One Edinburgh research worker has reported finding 217 organisms in rotting seaweed, of which 74 per cent had nitrogen-fixing properties—that is, were capable of producing nitrogen as food for plants.

In an earlier chapter we saw how soil bacteria could produce nitrogen for plant roots from seaweed, as well as certain substances which help to improve the physical structure of the soil. There is another way in which bacteria can help plants—by making available to them certain seaweed contents which are insoluble in soil water, and therefore unavailable to plant roots. These insoluble contents are carbohydrates which are broken down by the bacteria into organic acids. These acids then react with insoluble substances in the soil to produce soluble salts. These soluble salts (and trace elements with them) then become available for plant use. In this way it is possible, by feeding soil bacteria with the nitrogen, potassium, phosphorus, carbohydrates, trace elements, vitamins and other nutrients in seaweed, to help soil bacteria to feed agricultural and horticultural crops in turn. We think seaweed extract, in particular, has its dramatic effect on plant growth not only because it gives the plant direct nourishment, but also because it encourages soil bacteria to feed the plant, and thus helps it to make better use of nutrients potentially available to it.

*Pest deterrent*

The value of seaweed and seaweed products as pest deterrents has

5a. Right-hand runner bean seedling raised in compost without any base fertilizer. Fibre pot soaked in seaweed extract for 24 hours and then dried. Left-hand seedling in untreated pot and John Innes seed compost.

5b. The root systems of the two seedlings.

6a. Right-hand broad bean seedling raised in compost without any base fertilizer. Fibre pot soaked in seaweed extract for 24 hours and then dried. Left-hand seedling in untreated pot and John Innes compost.

6b. The root systems of the two plants.

been proved. The attacks of red spider mite on apples; of glasshouse red spider mite on chrysanthemums and cucumbers; of aphides on broad beans, sugar beet and strawberries; and of eelworm, tobacco mosaic, botrytis and other pests on a wide range of plants—these have all been reduced by the use of seaweed sprays or seaweed meal.

In trials with apple trees at Boreham in Essex, red spider mites on a seaweed-sprayed 20-leaf sample were reduced from 61 to 4, 100 to 12, 22 to 19 and 89 to 8. During the same period the mite population on unsprayed samples increased from 104 to 252, 77 to 302, 33 to 162 and 109 to 142. The trials took place in two orchards already infested with red spider mite, both newly planted with trees not more than three or four years old. It was therefore possible to ensure thorough spraying of each tree by pressurized knapsack sprayer. Spraying was uniform in all treatments, each tree being sprayed to 'foliage run-off'.

Twelve blocks of three trees each were used in each orchard. Four blocks were left untreated as controls, four were treated with 'Maxicrop' seaweed spray and four with a conventional acaricide. In the first orchard this was demeton-methyl, in the second, tetradifon.

It was decided, in the case of the seaweed-treated trees, to adopt usual commercial practice whereby extract is applied as a foliar feed in a sequence of sprays, at the rate of from 6 to 8 gallons an acre. Three such sprays, in a dilution of 1 in 50, were given at fortnightly intervals on the first site, and at weekly intervals on the second.

Red spider mites and eggs were counted seven times on each site. On the site sprayed fortnightly, the first count took place before spraying began on 15th June, the seventh after final spraying on 16th July. It was in the period between these dates that the reductions and increases earlier described occurred.

A reduction in the number of eggs also occurred in the seaweed-treated blocks, although this reduction was less than in the case of the mites. On the other hand, a large number of the eggs seemed to be dead. During the same period there was an increase in the number of eggs in the unsprayed blocks—an increase comparable to that which took place, in the same block, in the number of red spider mites.

At the second site, where spraying took place weekly, the first count was made before initial spraying on 16th July, the last after final spraying on 30th July. Here reductions in the number of red spider mites were from 62 to 21, 114 to 19, 79 to 9 and 46 to 10. During the same period the red spider mite population in the unsprayed blocks increased from 104 to 302, 93 to 192, 64 to 284 and 32 to 135. The number of eggs in one of the seaweed-treated blocks decreased, but in the other three increased, although at a much slower rate than in the untreated blocks. The weekly treatment was less successful than the fortnightly treatment because the overall period was shorter.

The interesting thing about these results is that there should be any reduction in the number of red spider mites, or eggs, at all. We intended seaweed extract to be used for feeding trees, and other plants, to stimulate their growth and yield. The fact that it helps plants to resist the attacks of pests and diseases was brought to our attention by those who use our products and those of other manufacturers. It was unexpected, and so far remains unexplained—although the fact that extract upsets the breeding cycle of aphides and other pests without killing their predators must be an important factor.

On the next page are the figures in tabulated form. Sample figures for trees sprayed with tetradifon and demeton-methyl are included for purposes of comparison.

These findings are to some extent confirmed by field trials in the United States, although in this case seaweed extract was added to an acaricide already in use. These trials took place with peaches at Ernella Orchards in New Jersey, and reductions were from a factor of 232 to one of 150. Our own seaweed extract, marketed in the United States under the name 'Sea-Born', was used.

These acaricide trials were conducted by Dr. Byrley F. Driggers, Professor Emeritus of Rutgers University, New Jersey, and a research specialist in entomology and economic zoology. Two ½-acre plots of mature Triogem peaches were tested. One was sprayed with 500 gallons of parathion-captan solution on four occasions. The other was given identical treatment, save that 2 quarts of seaweed extract were added to the total amount of spray used. When observations began on 20th and 28th May, adult red spider mites from over-wintered eggs were present on both plots. For seven successive weeks no mites were found on the seaweed plot, and only an occasional mite on the control plot. By 2nd September there were 232 adult mites on a 100-leaf sample from the control plot, and 150 on a 100-leaf sample from the seaweed plot.

Seaweed sprays have also reduced the number of glasshouse red spider mites on chrysanthemums at Knowlers' Nurseries, Much Hadham, Hertfordshire. Two plots, each measuring 150 ft. by 4 ft., were used in trials there. The control bed was given a soil treatment of 1 in 800 extract fortnightly, and sprayed with an acaricide on 27th

## TABLE 22

### Red spider mite on apples (1)

The effect, on an orchard infested with red spider mite, of spraying with 'Maxicrop' on 15th June, 29th June and 16th July 1964, and with demeton-methyl on 15th June and 29th June. All counts are for a 20-leaf sample.

| | *'Maxicrop'* | | | | | | | | *Unsprayed* | | | | | | | | *Demeton-methyl* | | | | | | | |
|---|---|---|---|---|---|---|---|---|---|---|---|---|---|---|---|---|---|---|---|---|---|---|---|---|
| | *M* | *E* | *M* | *E* | *M* | *E* | *M* | *E* | *M* | *E* | *M* | *E* | *M* | *E* | *M* | *E* | *M* | *E* | *M* | *E* | *M* | *E* | *M* | *E* |
| Pre-spray count | 61 | 53 | 100 | 66 | 22 | 61 | 89 | 43 | 104 | 105 | 77 | 62 | 33 | 78 | 109 | 31 | 52 | 49 | 77 | 135 | 25 | 140 | 35 | 69 |
| Final count | 4 | 51* | 12 | 35* | 19 | 61* | 8 | 27* | 252 | 215 | 302 | 251 | 162 | 115 | 142 | 166 | 3 | 6 | 9 | 11 | 5 | 18 | 9 | 15 |

* Most of these eggs seemed to be dead. M means mites, E means eggs.

## TABLE 23

### Red spider mite on apples (2)

The effect, on an orchard infested with red spider mite, of spraying with 'Maxicrop' on 16th July, 23rd July, 30th July, and with tetradifon on 16th July and 30th July. All counts are for a 20-leaf sample.

| | *'Maxicrop'* | | | | | | | | *Unsprayed* | | | | | | | | *Tetradifon* | | | | | | | |
|---|---|---|---|---|---|---|---|---|---|---|---|---|---|---|---|---|---|---|---|---|---|---|---|---|
| | *M* | *E* | *M* | *E* | *M* | *E* | *M* | *E* | *M* | *E* | *M* | *E* | *M* | *E* | *M* | *E* | *M* | *E* | *M* | *E* | *M* | *E* | *M* | *E* |
| Pre-spray count | 62 | 89 | 114 | 53 | 79 | 133 | 46 | 81 | 104 | 100 | 93 | 85 | 64 | 75 | 32 | 101 | 74 | 103 | 69 | 53 | 155 | 162 | 67 | 83 |
| Final count | 21 | 58 | 19 | 132 | 9 | 151 | 10 | 111 | 302 | 245 | 192 | 104 | 284 | 183 | 135 | 103 | 2 | 31 | 3 | 20 | 9 | 41 | 3 | 61 |

Egg counts in the case of tetradifon do not reflect actual control effect, full effect being shown three or more weeks after spraying.
M means mites, E means eggs

November, 10th January and 22nd January. The experimental plot was given a 1 in 200 foliar seaweed spray as well, at fortnightly intervals. After fourteen weeks there were just over 30 mites and 25 eggs on a 50-leaf sample taken from the seaweed-sprayed plot, about 185 mites and 90 eggs on a similar sample taken from the acaricide-sprayed plot. Fortnightly samplings during the period showed that the number of mites on the control bed was continually increasing, while the number on the seaweed-sprayed bed was being progressively held. The cost of the seaweed treatment was only a fraction of that of the acaricide. Here are the results in graph form.

TABLE 24

Red spider mite on chrysanthemums

Graphs showing the effect on mite and egg numbers of foliar spraying with 1 in 200 solution of seaweed extract at fortnightly intervals in glasshouses at Much Hadham, Hertfordshire. Numbers are for 50-leaf sample.

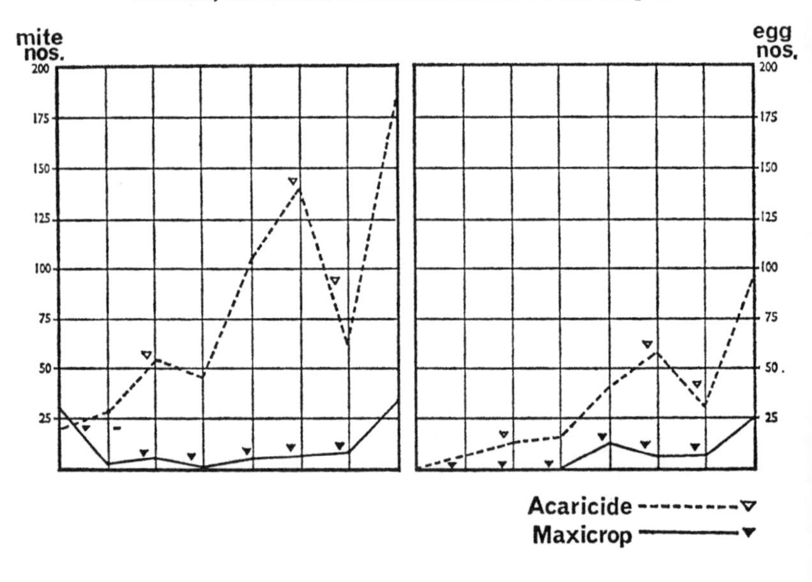


In trials with cucumbers under glass at the same nurseries, it was found that seaweed extract was slightly more effective against glasshouse red spider mite than standard commercial acaricide—and cost half as much. Plants treated with seaweed extract produced a 14 per cent heavier crop—1,178 trays compared with 1,027 trays—and it was also obvious that the seaweed-treated plants suffered much less from botrytis.

Two adjoining glasshouses without a dividing wall were used in the trials. They were of exactly the same size and type, and each contained 300 plants. One house was given 50 gallons of 1 in 200 seaweed extract solution as a foliar spray every ten days. The other was sprayed with a standard acaricide at the same intervals. Spraying began on 4th April, although we know now that it should begin at seedling stage.

The degree of infestation in each house was assessed at intervals, and the results recorded according to an accepted scale: 0, for no damage; 1, for incipient damage, with one or two ½-in. feeding patches; 2, for feeding patches tending to coalesce, but with only two-fifths of the leaf affected; 3, for two-thirds of the leaf having feeding marks showing as chlorotic patches; and so on. This scale is also related to the number of mites found on each square inch of leaf.

Routine assessments of the two glasshouses gave the following results, with danger levels in italic numerals.

TABLE 25

Red spider mite on cucumbers

Effect of 1 in 200 solution of seaweed extract, and of a solution of standard acaricide, sprayed every ten days on glasshouse cucumbers infested with red spider mite at Much Hadham. Degree of infestation assessed, with danger levels in italics.

| | *Seaweed extract* | *Standard acaricide* |
|---|---|---|
| April 23rd | 0·08 | 0·51 |
| May 11th | 0·39 | 0·86 |
| May 26th | 0·84 | 1·44 |
| June 11th | 1·13 | 1·33 |
| June 25th | 1·03 | 1·37 |
| July 20th | 1·25 | *1·76* |
| August 3rd | 1·31 | *1·87* |
| August 20th | *2·08* | *2·35* |

There is evidence that seaweed sprays discourage aphides, in particular the black bean aphis. This carries the disease known as 'virus yellows', which can reduce yield of sugar beet by as much as 50 per cent.

In trials at Cambridge, black bean aphides caged on broad bean plants treated with a seaweed spray produced fewer young than those caged on similar untreated plants. Whenever the cages were examined, it was found that more aphides were wandering restlessly about the seaweed-treated leaves than about the others—and it is known that restlessness in adult aphides results in a lower rate of reproduction.

These observations have been confirmed by the use, on sugar beet plants, of a cage which gives aphides a choice between treated and untreated plants. (Cages are flat containers an inch or two across. They are clipped on infested leaves so that aphides, and other insects, may be restricted to a particular plant and observed there at intervals. Those used on the bean plants restricted aphides either to treated or to untreated plants. Those used on the sugar beet offered them a choice.)

In the sugar beet trials, thirty young and active winged black bean aphides were released in the 'choice' cage, and observed every twenty minutes. It was then found that over a period of five hours, four times as many aphides settled on untreated as on treated plants.

We have also received reports from the Vale of Evesham that brussels sprouts sprayed with demeton-methyl and seaweed extract remained free from aphis attack for six weeks, while those sprayed with demeton-methyl alone become infested after three.

Similar research has taken place in the United States. At Rosenhayn in New Jersey, for example, the effect of seaweed spray on aphides which attack strawberries has been studied. Seaweed extract, applied in solution at the rate of 1 gallon to the acre, so reduced the fecundity of the aphides that as a result infestation dropped by 56 per cent, compared with kills of 92 and 96 per cent achieved by use of two chemical aphicides. The seaweed treatment did not reduce the number of leaves infested, which suggests that its effect was to reduce the aphides' capacity for reproduction rather than to kill them.

The extract used in these Rosenhayn trials was 'Sea Magic' (Chase Organics Ltd., Shepperton, Middlesex), an extract of 80 per cent Norwegian seaweed mixed with 20 per cent land plants. A new formula known as 'Sea Magic' 131, made available for experimental purposes in 1963, was used in the trials.

Seaweed has shown encouraging results against soil pests such as eelworms. The States experimental station in the Channel Island of Jersey has reported that tomatoes manured with seaweed have cropped well in soil known to be infested with eelworm. This station has also reported consistently good crops of tomatoes from eelworm-infested soil treated with about half the recommended quantity of the soil sterilizer dichloropropane and dichloropropylene, known commercially as D-D, but only as long as heavy dressings of seaweed are given as the sole fertilizer.

Verticillium wilt in carnations has been much reduced by the use of seaweed meal at Kobe Nurseries, Halstead Hill, Cheshunt, Hertfordshire. Mr. A. G. Roberts has two houses there, each containing four carnation beds, 4 ft. wide by 150 ft. long. He had no time to steam the beds before putting in plants ready for cutting in the summer of 1966, but he did try putting 1 cwt. of seaweed meal on one bed in each house. And in May 1966, while carnations in the other beds were attacked by verticillium wilt, those in the two beds given seaweed meal were free of the disease.

The fungus which causes foot rot in beans has been reduced in a heavily infected soil from an arbitrary index of 66 in untreated soil to one of 18 in treated soil by adding laminarin, a starch of brown seaweed. This was reported in 1963 by Dr. Ralph Mitchell of the Laboratory of Soil Microbiology at Cornell University in the United States. In other trials Dr. Mitchell found that the severity of several soil-borne fungus diseases of plants was reduced when chitin or laminarin, both seaweed products, was added to the soil. The plot in which the experiments took place, which was planted with red kidney beans, had been heavily infected with fungus for over forty years; but chitin and laminarin stimulated the soil bacteria which (to paraphrase Dr. Mitchell) then ate the offending fungus.

Similar results were obtained with radishes suffering from fungus attack. Chitin reduced attack from a factor of 63 with no treatment, to one of 43, laminarin from 63 to 40. In another infected plot, reductions were from a factor of 59 to 37 when chitin was given, and to 25 when laminarin was given. (Factors were based on a scale varying from 0, for no disease, to 100, for plant killed.)

Trials by Charles Booth, a consulting entomologist of Cambridge, have also shown that plants sprayed with liquid extract are less likely to be attacked by mildew. When sixty 'Snowball' turnip plants in 3-in. pots were arranged in matching pairs in a greenhouse, in conditions favourable to powdery mildew, only 15 per cent of the leaf surface of plants treated with extract was attacked, compared with 85 per cent of untreated plants. The treated plants—one in each pair—were watered weekly with a 1 in 120 solution; and the experiment ended when the plants most affected were completely covered with fungus.

Trials have also taken place at Cambridge to test the value of extract against botrytis in strawberries. In the first, thirty plants of 'Cambridge Favourite' in matching pairs were used. These had been forced, and were already attacked by botrytis. (Forced strawberries are always more subject to this disease.) They were then forced again in wet and warm conditions, and one plant in each pair watered with a 1 in 120 solution each week until fruiting. The plants were fertilized daily with a camel-hair brush, and the fruit picked as it ripened. At the end of the season, 4·6 per cent of the treated berries were infected, compared with 22·5 per cent of the untreated.

In the following year the same plants, which had stood outside during the summer and been watered every three months with John Innes liquid feed, were brought into the greenhouse when flower buds developed. After being treated as before, treated berries with botrytis numbered 4·8 per cent, untreated 15·4 per cent.

In a third strawberry trial, 100 rooted runners were potted in the autumn with John Innes compost No. 1, and forced until the fruit was set. Humidity and warmth were then increased to encourage botrytis further. After being treated as in the two previous trials, seaweed-treated berries infected with botrytis numbered only 1·7 per cent, while untreated berries with infection totalled 12·9 per cent.

In all these strawberry trials, treated plants produced more fruit—14·8, 31·9 and 28·0 per cent more respectively.

The fact that seaweed extract is used to discourage tobacco mosaic in tomatoes came to our notice in an unusual way. When we found we were sending considerable quantities of our extract to Guernsey we at once went into the reasons for this; seaweed fresh from the sea is used so widely in the Channel Islands that it seemed unlikely that growers there could have any use for liquid extract. We then discovered, to our surprise, that it was being used against tobacco mosaic in tomatoes. At the first sign of the virus, growers spray their glasshouses with a strong solution—say 50-50—of the extract in water. This does not eradicate the disease, but it does keep it within bounds, and allows a reasonable crop to be gathered.

One seed-borne fungus disease which has responded to seaweed treatment is that which attacks leeks in the north-east of England. It causes particular trouble to those who vie with each other in growing the largest specimens for competition. One grower at least has overcome this problem by soaking seed in a weak solution of seaweed extract for twenty-four hours. Dried on blotting paper and sown a day or two later, the seed now germinates without sign of fungus disease.

Botrytis and damping-off among lettuces in cold houses in the Blackpool area have been kept at bay by spraying with seaweed extract. Seaweed spraying has also given promising results with rust fungus in dwarf runner beans, and leaf curl in peaches. As to the first, I have little to go on save the comments of customers. As to the second, not a summer goes by but someone growing peaches writes to tell us that seaweed extract has cured leaf curl, a complaint often regarded as incurable.

Our experiences with foliar sprays based on seaweed, together with the experiences and observations of others, lead to two general conclusions: the resistance to pests and diseases given by these foliar sprays becomes more and more noticeable after two or three years' use; and, within limits, the weaker the solution, the stronger the effect. Certainly the same amount of liquid seaweed extract produces a greater effect if given in a series of weak solutions, rather than in the form of one or two stronger solutions.

It is difficult to explain these effects. Seaweed doubtless increases the disease-resisting qualities of plants by helping them to become healthy and vigorous. I would go further, and say that the auxins it contains discourage fungus and virus diseases; and that through these, and other constituents, seaweed encourages soil bacteria to produce natural antibiotics for plant and soil.

The resistance which seaweed seems to give plants against mites and aphides cannot be explained on these grounds. I can only point out that aphides have a well-developed sense of taste. Perhaps they dislike the taste and smell of leaves which have been sprayed with extract. Seaweed also produces a thicker cuticle, so a sap-sucking insect must work harder. It also tends to reduce water content, so that less sap is available. The trace elements in seaweed may also be a factor in discouraging mites and aphides. Research workers at Oregon State University in the United States have found that the fecundity of spider mites can be reduced as much as 90 per cent if they are sprayed with a solution of the metal chelates of iron, manganese, zinc and magnesium. When these chelates, supplied by Geigy Company Ltd., were used on beans, strawberries and sweet potatoes, the sprays reduced the fecundity of two-spotted spider mite (a similar pest to red spider mite) by as much as 90 per cent. Spider mites and aphides on hops have also been inhibited for several weeks by foliar sprays containing chelates of the same metals.

### *Frost-hardiness*

Trials at Clemson College have confirmed what is a matter of experience among nurserymen in this country, that seaweed helps plants to resist frost. Trays of tomato plants grown in soil to which

1 in 30 of seaweed meal had been added withstood temperatures of 29° F., although the control plants were killed. Tomato plants grown in greenhouse plots treated with seaweed resisted two successive frosts when planted out, although the first frost severely injured plants raised without seaweed. Citrus plants were found to be much more resistant to cold when sprayed with seaweed extract.

It is already known that vitamins help plants to grow at abnormal temperatures. H. J. Ketellapper at the Californian Institute of Technology has found that at abnormally low temperatures, nicotinic acid stimulates the growth of tomato plants, a mixture of B vitamins that of purple Mexican aster, and a mixture of ribosides that of eggplant. It is therefore probable that fresh seaweed, and seaweed meal, increase the cold resistance of plants because of the vitamins they contain. This, however, does not explain why tomatoes, celery and citrus fruits sprayed with seaweed extract withstand frost better than those not sprayed, for the vitamins in seaweed are the only effective constituent removed by hydrolysis.

*Shelf life*

Some of the most dramatic results of treating fruit trees with seaweed products are seen in the increased shelf life of fruit. At Clemson College, for example, ripe peaches from trees which had been sprayed with extract had a shelf life four times longer than those from unsprayed trees, as the following table shows. The tests took place on six plots in a 20-acre orchard. The first of the columns shows the state of the fruit four days after harvesting, the second five days after, and so on. All the figures in the table refer to rotten fruit. They are brought to daily totals underneath each group of plots, with overall totals of rotten fruit in the last column to the right.

Similar evidence comes from New Jersey in the United States, where the manager of Ernella Orchards in Warren County—to which reference has already been made—found a remarkable improvement in the shelf life of early ripening peaches after 1 quart of seaweed extract had been added to 500 gallons of his regular parathion-captan spray. After one single application of this spray, between 75 and 100 customers—including housewives, fruiterers, the owners of wayside fruit stalls, and one wholesaler—asked him what he had done to make his peaches last so much longer. He decided to test the value of seaweed extract further, and carried out field trials which were reported in the May 1964 issue of the United States journal, *Horticultural News.*

TABLE 26

Shelf life of peaches

Fruit from plots at Clemson College sprayed with seaweed extract compared with fruit from unsprayed plots

| | | *Number of rotten fruit from 4 to 12 days after picking* | | | | | | | | | *Totals of rotten fruit* |
|---|---|---|---|---|---|---|---|---|---|---|---|
| | | *4* | *5* | *6* | *7* | *8* | *9* | *10* | *11* | *12* | |
| | A | 6 | 2 | 1 | 0 | 0 | 2 | 0 | 4 | 0 | 15 |
| Untreated | B | 2 | 5 | 1 | 0 | 0 | 0 | 0 | 3 | 0 | 11 |
| plots | C | 3 | 5 | 1 | 0 | 0 | 0 | 0 | 2 | 0 | 11 |
| | Total | 11 | 12 | 3 | 0 | 0 | 2 | 0 | 9 | 0 | 37 |
| | D | 0 | 0 | 1 | 1 | 0 | 0 | 0 | 0 | 0 | 2 |
| Treated | E | 0 | 0 | 0 | 0 | 0 | 1 | 0 | 1 | 0 | 2 |
| plots | F | 3 | 0 | 0 | 0 | 0 | 0 | 0 | 0 | 1 | 4 |
| | Total | 3 | 0 | 1 | 1 | 0 | 1 | 0 | 1 | 1 | 8 |

In the trial, which took place in 1962, the concentration was increased to 1 quart in 100 gallons, which was sprayed four times. Six bushels of peaches were collected at random from each of the sprayed plots on 12th, 16th and 21st August, and all put into cold storage after picking. They were brought out on 28th August, and put in an open shed at ordinary temperature, side by side with peaches similarly treated in every way, save that no seaweed extract had been given them.

When the baskets were sorted on 3rd and 6th September, and peaches showing sign of rot discarded, twice as many peaches from the control plots had to be thrown away. Of the 961 peaches from the seaweed plots, 14·6 per cent showed rot, of the 874 from the control plot, 32·6.

*Quality*

The effect of seaweed on the quality of agricultural and horticultural produce is a difficult thing to deal with—not that the improvements are not there, but because to detail them becomes tedious. So many commercial interests make inflated claims for the quality of inferior food, clothes, shoes, drink and other products, that for a seaweed producer to make such claims for his products may make the reader suspicious. I can only say that I believe seaweed products do stiffen flower and cereal stems; do make lawns springy underfoot; do enhance the colour of most flowers, particularly freesias, begonias and chrysanthemums—and do make these, and other flowers, last longer; do give a deeper colour to carrots; do give strawberries a better and sharper flavour; do produce short-jointed, vigorous tomato plants, with a low bottom truss, whose fruit has better flavour and keeping quality, and an increased resistance to disease; and do increase the quality, and quantity, of flower perfume.

I will refer to two cases only: that of the British Embassy lawn at Washington in the United States; and that of the pinks used to decorate the tables at the Colchester oyster feast.

In the spring of 1959, 4 lb. of dehydrated liquid seaweed extract were reconstituted and used on a measured part of the Embassy lawn; and when in 1960 this patch came up stronger and greener than the rest, we received an urgent request for a further 8 lb. to be sent as quickly as possible, so that the whole of the lawn might show up even better on an important social occasion.

As far as the oyster feast is concerned, all I can do is repeat the statement of Mrs. Desmond Underwood, who grows the pinks for the feast tables—that these flowers have a much stronger perfume when given seaweed rather than artificial fertilizer.

# 9

# HOW SEAWEED PRODUCTS ARE APPLIED

Seaweed meal used as soil conditioner should be spread on the ground after autumn cultivations. If it is left on the surface of the ground, rain, worms and bacteria will do the rest. Only in the case of some heavy clay soils may it be necessary to work it in. With such stiff clays, dressings of from 6–8 oz. a square yard may have to be given every year for two or three years before there is obvious improvement. Clay loams respond sooner than stiff clays; 6–8 oz. of seaweed meal to the square yard will give a workable seedbed ½ in. deep in six to eight weeks. Two annual applications will give a 6-in. deep seedbed in a couple of seasons.

These instructions are for private gardeners. Commercial growers would not tackle such stiff clay soils, and farmers forced to work them would find seaweed too expensive as a soil conditioner. Nor, as we have seen, do these recommendations apply in the United States, where less seaweed meal (and seaweed extract) need be used to give the same effect.

So far, too, we have been considering only one aspect of soil conditioning: that of making soil workable. In the case of sandy soils, such as are found over large areas of Surrey, it is not a question of ease of working, but of improving the water-holding qualities of the soil—and, with it, the ability of the soil to retain nutrients held in solution by ground water. These sandy soils are warm and dry, and sometimes highly productive; but because of their failure to hold soil nutrients in solution they need continued, regular feeding. Their fertility and water-retaining properties can, however, be improved by annual applications of seaweed meal and peat given as generously as possible. With the hungriest of these soils, even 8 oz. of seaweed meal to the square yard may not be enough.

With chalky soils we move even further away from pure soil conditioning—for here it is a matter of trace element unavailability. Chalky soils will, of course, have their water-retaining qualities improved by seaweed, for which purpose it may be added at the rate of 4–6 oz. a square yard. But improvements in this direction may be masked by more dramatic improvements resulting from the freeing of trace elements; and for this purpose the weight of meal used may be less.

Other soils such as loam, fen peats and silt are so free-working that seaweed is unnecessary as a soil conditioner. Fen peats and silts are also too fertile to need seaweed meal as fertilizer, although it may be offered, where necessary, in answer to trace element deficiency. But even here deficiencies will probably be more efficiently, and cheaply, countered by foliar spraying of seaweed extract.

It will be understood that the amounts of seaweed meal I have suggested for conditioning and fertilizing garden soil may be varied in the light of experience. In all cases, too, it will be found that results are cumulative. There may be few signs of improvement in the first year or so: there will be many in the third and fourth.

Seaweed meal should not be used in pot mixtures unless it has been composted, or otherwise specially prepared. The method of composting invented by Mr. Chilcott is described by him in a later chapter. It is the most effective way of reducing the 'dead time' which follows the application of pure seaweed or seaweed meal, but it can be used on a large scale only. Any compost heap containing less than 1 cwt. of seaweed meal would not retain its heat in the conditions he outlines.

There are two other ways of treating seaweed meal intended for potting. Mixtures can be made up one month before they are needed; or meal may be added to peat in such proportions, and so placed in the pot that by the time the plant's roots encounter the meal, composting has already begun. This is best done by mixing a teaspoonful of seaweed meal with half a handful of wet peat and putting it at the bottom of the pot, with nothing beneath it but the crock over the hole—or, in the case of a plastic pot, the holes themselves. The pot is then filled with whatever compost is thought advisable. Then, when the roots penetrate the peat, composting of the seaweed will have already begun.

It is only right to point out, however, that with moderately rich loams it is as effective, and cheaper, to use liquid seaweed extract in making up potting compost. All that is necessary here is to soak the peat in extract before mixing. With poor and sandy loams, however, meal should be used as earlier described.

If meal is needed for general garden use it can be composted by the methods outlined above, although in the ordinary way such speeding-up of natural processes, necessary with pot culture, is unnecessary out of doors. On the other hand, seaweed products can help to decompose ordinary garden refuse. So used, they feed the bacteria of decay in the same way that, in the form of agar, they feed the bacteria in laboratory cultures.

Compost heaps are bacterial bonfires: activators such as I have described are the papers and sticks which start the fire. To do this they must supply readily available food for the bacteria, which then multiply and cause a swift rise in temperature—and in the process decompose first the activator, then the heap. The heat they produce 'cooks' weed seeds and the roots of potatoes, and destroys the spores of plant diseases. But in order to kill rust spores, and the root and stem eelworm of phlox and daffodils, the temperature must rise well above 115° F.; otherwise, material which has not risen to this temperature will produce a forest of seedling weeds when spread on the soil.

Two compost activators based on seaweed, 'Alginure' and 'Marinure', have been tested by the Henry Doubleday Research Association at Bocking in Essex, and the association's reports of the trial are worth describing in detail. In the first, which took place in 1961 and 1962 with 'Alginure', a New Zealand box was used; in the second, which took place in 1966 with 'Marinure', two New Zealand boxes and one Bocking box were used.

Boxes are used for composting because wind can lower the temperature on one side of an open heap and slow down decomposition there. The heap then decays unevenly as a result. Boxes hold the heat, spread it evenly throughout the mass, and ensure that decomposition continues to the edge of the heap. New Zealand boxes are topless and bottomless boxes with two or three compartments. They have sliding boards at the front, which can be moved for shovelling out the finished product. Those in the trial had two double rows of bricks in the bottom of each compartment. These protruded under the removable board fronts, and so allowed a stream of air to pass up through the heap. Such a draught serves the same purpose in a bacterial bonfire as in a real fire. Bocking boxes are pens made of newspaper sandwiched between two layers of wire netting, the whole being secured with rough stitching. Those used in the trial also had two double rows of bricks at the bottom to provide a constant supply of air.

'Alginure', the first activator tested, is a seaweed jelly containing a culture of bacteria. It is sold in a giant 'toothpaste' tube. The bacteria in the tube are dormant while the top is in place, and the same tube was in fact used for trials in 1961 and 1962.

The association's report describes the construction of the heaps and the method of applying the activator, adding the following table to show how the temperature in the treated heap was increased by the use of 'Alginure'.

TABLE 27

Composting with 'Alginure'

This table shows the effect of 'Alginure' on the temperature of a compost heap

| *With 'Alginure' Heap finished August 25th* | *Degrees Fahrenheit* | *Without 'Alginure' Heap finished September 7th* | *Degrees Fahrenheit* |
|---|---|---|---|
| August 27th | 80 | September 8th | 82 |
| 28th | 150 | 9th | 74 |
| 29th | 150 | 10th | 70 |
| 30th | 158 | 11th | 82 |
| 31st | 130 | 12th | 76 |
| September 1st | 125 | 13th | 110 |
| 2nd | 108 | 14th | 94 |
| 3rd | 70 | 15th | 80 |
| 4th | 68 | 16th | 62 |
| 5th | 72 | 17th | 68 |
| 6th | 72 | 18th | 58 |
| 7th | 76 | 19th | 64 |
| 8th | 66 | 20th | 60 |

The report concludes: ' "Alginure" is the simplest compost activator, needing no mixing, lime layers, soil layers or scarce manure, and its rapid heating in spring and summer makes it useful where domestic refuse is included in the middle of the heap, for temperatures like 150 Fahrenheit destroy smells and make fly maggots impossible. There should be no problems with a good air supply, and though the tube may seem relatively expensive it is only so for those who squeeze out too much, and each one should make the four hundredweights of finished compost claimed by the makers easily. As the heap sinks it is possible to go on adding till each compartment in the box finishes as a solid cube of good stuff for autumn or spring use. Average making time in summer is seven weeks; the last heap made in October of cut-down rubbish and haulm is best left till spring and it is advisable

7. John Cooke of Funtington, Sussex, with his prize-winning potato plant. He soaked the chitted seed in seaweed extract before planting.

8. *Fucus* seaweeds growing on Scottish coast. Bass Rock in background. By courtesy of Institute of Seaweed Research, Inveresk.

to cover it down with something to keep off heavy rain and snow from December, as these can wash out plant foods.'

The first of two tests with 'Marinure' dried seaweed powder took place in Bocking boxes, compost in one box being treated with 'Marinure', that in the other with the 'Alginure' jelly activator just described. The temperature in both heaps rose to 140° F., which was regarded as excellent in heaps which contained a good deal of cold soil. The seaweed meal heap took forty-eight hours to begin heating up, although the jelly heap began working next day. 'Both had the reliable heating ability typical of seaweed activators, as distinct from herbal or bacterial culture types which can fail for the unskilled,' the report commented.

In the second 'Marinure' test three heaps were made, one of potato and tomato haulm; one of weeds, sunflower stems and comfrey; and one of weeds, comfrey and grass. The first two were treated with 'Marinure' seaweed meal, the third with 'Alginure' jelly. The potato and tomato haulm heap, which contained some potato haulm showing blight, was made to see how 'Marinure' worked with sappy material containing no soil. As table 28 shows, it reached a temperature of 130° F. in four days, and maintained that temperature for five days. Since the spores of potato blight die between 120° and 130°, this was enough to deal with blight in the heap.

The second heap containing weeds, sunflower stems and comfrey was designed to see how a seaweed meal compost activator tackled tough material, including tall sunflower stems cut in 2 ft. lengths, and split down the middle. In the event a temperature of 120° and over was recorded for thirteen days. 'A very fine achievement,' the report comments.

On the next page are the temperatures in full.

The report states in conclusion: 'The tests, with heaps made by gardeners who had never made any before, show that "Marinure" makes good compost as simply as the chemical powders sold in tins and rather more cheaply, because it is bought at a bulk fertilizer price and can be stored indefinitely. It is as effective as a seaweed jelly, and it is suggested that a tin with a well-fitting lid and a measure inside to hold two ounces should be kept by the heap for use as soon as an eight-inch-thick layer of weeds and rubbish has been gathered. The better decay is probably due to the more even distribution, so scatter the measureful as thinly as possible.

'No tests have been made with straw or dry woody materials, but these should be mixed in with garden rubbish, and should decay if not included in too great a quantity. Kitchen wastes, which can include not more than ten per cent of paper, can go in the middle of the heap and the rapid heating will break them down without flies or smell.

TABLE 28

Composting with 'Marinure'

This table shows the effect of 'Marinure' and 'Alginure' on the temperature of compost heaps

| *Potato and tomato haulm treated with 'Marinure'* | *Degrees Fahrenheit* | *Weeds, sunflower stems and comfrey treated with 'Marinure'* | *Degrees Fahrenheit* | *Weeds, comfrey and grass treated with 'Alginure'* | *Degrees Fahrenheit* |
|---|---|---|---|---|---|
| August 31st | 90 | October 14th | 150 | October 14th | 130 |
| September 1st | 104 | 15th | 160 | 15th | 130 |
| 2nd | 106 | 16th | 145 | 16th | 120 |
| 3rd | 130 | 17th | 140 | 17th | 120 |
| 4th | 155 | 18th | 145 | 18th | 100 |
| 5th | 140 | 19th | 120 | 18th | 95 |
| 6th | 150 | 20th | 160 | 19th | 95 |
| 7th | 140 | 21st | 155 | 20th | 90 |
| 8th | 130 | 22nd | 158 | 21st | 88 |
| 9th | 120 | 23rd | 140 | 22nd | 80 |
| 10th | 110 | 24th | 130 | 23rd | 80 |
| 11th | 110 | 25th | 125 | 24th | 80 |
| 12th | 110 | 26th | 120 | 25th | 80 |
| 13th | 90 | 27th | 102 | | |

'Those whose compost heaps harbour woodlice and slugs, and whose compost never heats enough to prevent thick germination of weed seeds, should check their air supply from below, and the activator. It is highly unlikely that their garden lacks the right bacteria—they are far more likely to lack a good activator. "Marinure" is one of the best.'

Since 'Marinure' is whole seaweed dried and ground, with nothing added and nothing taken away, it is true to say that any standard seaweed meal used as a compost activator will produce comparable results.

So far I have dealt with seaweed meal only. I must now say something about the use of liquid seaweed extract, which has a greater variety of applications.

Liquid seaweed extract is compatible with all other sprays. It works well in trickle systems, because it neither corrodes nor crystallizes out of solution. Nor does it need further agitating once it has been mixed.

And if, for the moment, the mixing and spraying of extract may be considered without relation to the needs of the plant, dilution rates may be decided, as a matter of convenience, according to the type of sprayer used. (In practice, of course, while dilution rates can be varied enormously without damaging a crop, stronger solutions may retard the plant temporarily, and weaker solutions may produce responses quite unrelated to their strength. These are not matters which concern mixing and spraying viewed simply as a physical operation, however, and are dealt with later.)

Solutions of extract can be made up weeks before they are used. If the mixture contains herbicide or other chemical, however, it should be applied fresh, and not left in the spray tank overnight. If mixed with other spray material, the extract can be added neat or diluted.

The simplest dilutor to use is the venturi type, of which a number are made in sizes to fit ½, ¾, 1 and 1¼ in. bore hose. (Dilutors prepare spray solution by adding a small amount of one liquid to another—usually, but not always, of liquid concentrate to water. A venturi will not work against a back pressure, and is for use with open-ended hose only, and that not more than 50 ft. long. It is not recommended for use with trickle feeders or spraylines. These, when used, should be washed out regularly, or slime will form in the pipes and nozzles.) The dilutors we recommend for all-round use are the Keylutor (27 St. Albans Crescent, Woodford Green, Essex) and the Cameron (Cameron Irrigation Company Limited, Harwood Road Industrial Estate, Littlehampton, Sussex). They are available from 1 gallon capacity for gardeners up to 5 gallon capacity for growers.

I spoke earlier of dilution rates from a purely operational point of view. From the point of view of good husbandry, which is of course paramount, our own liquid extract, 'Maxicrop', and 'Maxicrop' with chelated iron, can both be given without ill-effect to mature plants, but not to sappy plants or seedlings, with equal quantities of water. 'Maxicrop'-with-nitrogen, however, should be in a dilution of at least 1 in 8 for low-volume spraying, and of 1 in 12 for high-volume spraying. In fact much weaker solutions than these are often to be recommended, particularly when applied at correspondingly smaller intervals. Applications—or so growers have told me—seem to be more effective when given in the morning, when the leaf stomata are open. The leaf stain which may appear on a plant immediately after it has been sprayed is harmless. It will disappear after three or four days.

Foliar spraying is only one of the ways in which seaweed extract may be used. Seeds, and the bulbils which are produced by some lilies, will germinate quicker if soaked in a 1 in 400 solution for twenty-four hours before sowing. If seeds are too small to handle, the compost in which they are sown may be soaked in a solution of the same strength. If the compost is home made, then the peat should be soaked in a 1 in 300 solution before being mixed with the loam and sand.

This technique of soaking compost before seeding may be used with seeds of any size; but where germination is of importance, soaking the seed itself is the most certain way of encouraging it. If fibre pots are used, soaking the pot in solution will avoid de-nitrification resulting from decomposition of the paper. (This is the same nitrogen-robbing process which takes place when straw, seaweed, or any un-composted plant residue is added to the soil.)

Cuttings, and the compost they are planted in, should be soaked in a 1 in 300 seaweed solution for twenty-four hours before setting; and thereafter given a weekly dose at the same strength. Soft or hard wood cuttings intended for the production of shrubs and fruit bushes should be soaked in a 1 in 300 solution for twenty-four hours before being put in the rooting medium. Thereafter the rooting medium should be watered with a 1 in 400 solution.

When bedding plants need holding back, dosing once with a strong solution of (say) 1 in 50 will keep them sturdy, healthy and short legged, and help them to get away quickly once they have been planted out.

If the soil in the bed is dry, it should be watered once with 1 in 100 solution. If wet, the plants should be put in solution, two to six hours before being transplanted. They will then be saved the flagging which root disturbance sometimes causes transplants; and will often grow away immediately.

We have found that seaweed solution, and commercial rooting compounds such as indolyl-acetic acid, both produce good rooting systems. But seaweed-treated cuttings seem to get away much quicker and, unlike those treated with indolyl-acetic acid, never have a set-back. Perhaps this results from the fact that while indolyl-acetic acid encourages cell elongation, the auxins in seaweed encourage cell division also.

Where transplanting takes place with mature objects such as fair-sized bushes or trees, the compost in the planting hole should be soaked with 1 in 100 solution. Once the plant is in position, the rooting area should be watered with the same solution. The easiest way to do this is to dig two or three quart-sized holes over the rooting area, and fill each with solution. A foliar spray should be given in April or May with the same solution.

When layering, the point where the layer is put in should be watered with 1 in 300 solution. With air layering, the sphagnum moss used may be soaked in solution of the same strength. (Air layering is a method of propagating plants—among them certain types of holly, lilac, magnolia, azalea and clematis—cuttings of which will not readily strike roots of their own accord. If cuttings are taken from these plants and put in the soil they usually die. But a rooted cutting can be prepared by removing a slim ring of bark $\frac{1}{4}$ in. wide all round one of the branches or stalks of the plant, or by making a tongue-cut, kept open with a matchstick, at the same point. If these are packed round with damp sphagnum moss tied in place, and covered with a polythene sheet secured at each end with adhesive tape, then roots will appear at the cut. If the branch or stalk is cut off immediately below this point, and the cutting potted, it will grow into a new plant.)

Some plants, including African violets, are propagated from leaf cuttings. They produce new plants from the base of the leaf. Others, including begonias, are propagated from splits made in the main rib of the leaf. Yet others, like sansevierias, are propagated from short pieces of leaf. Sansevierias have long, thin, sword-like leaves. If these are split into 2 in. lengths they will root when set in compost. In all these cases, soaking leaves and leaf sections may be a fiddling job; it is easier to soak the rooting medium in a 1 in 300 solution before pegging down leaf or leaf section—an operation, incidentally, for which I find hairpins of some service.

Bulbs will profit if fibre, soil or forcing boxes are soaked in a 1 in 300 solution before planting out. Bulbs to be kept for another season should have their leaves watered and, if possible, sprayed with 1 in 200 solution once flowering is over. This helps the bulb to store more food and to become more vigorous for the next season's flowering.

Once a bulb has begun flowering, little is to be gained by offering it any nutrient, seaweed included. But if the atmosphere in which the plant is growing encourages mildew, spraying or watering with 1 in 50 solution will help to inhibit it, particularly in the case of tulips.

Flowers which grow from corms—such as freesias, gladioli, crocuses and montbretias—produce more corms for further planting if they, or the compost in which they grow, are sprayed or watered with a 1 in 100 solution of seaweed extract. This is particularly effective with gladioli.

More detailed instructions for the use of seaweed extract in horticulture are given, for individual plants, in Appendix A.

# 10

# SEAWEED MEAL AS FEEDINGSTUFF

Animals have eaten seaweed in its natural state for as long as can be remembered. At times grazing animals turn to seaweed because of a shortage of other foods, particularly in a hard winter; but both domestic and wild animals will eat seaweed in high summer, even when ample supplies of grass and other herbage are available. I myself have seen sheep and cows nibble at seaweed growing on rocky summer beaches, and wade up to their bellies in the sea to graze seaweed floating on the water.

The fact that farm animals take to seaweed is also confirmed by the names given some of the weeds they eat. A little bunchy seaweed, the richest in fat of any, which grows near the top of the tide in the Orkneys, is known as 'calf weed'. It bears the same name, or its Spanish counterpart, on the Atlantic shores of Spain. In Norway, *Ascophyllum nodosum* is known as 'pig weed', and *Alaria esculenta* as 'cow weed'.

Sheep are probably the biggest seaweed eaters, although cattle, horses, pigs and goats all eat it—witness the names 'cow tang' in the Hebrides, and 'pig weed' on the Baltic island of Gotland. *Rhodymenia palmata*, which is known as 'cow seaweed' in the Roscoff area of Brittany, is known in Norway and Lapland as 'horse seaweed'. Contemporary records speak of the Greeks feeding seaweed to cattle in the first century B.C.; and reports of animals choosing to eat seaweed come from Iceland, the islands of the Behring Straits, from Norway, Finland, France, the Faroes, New Zealand, and all coastal areas of Scotland and Ireland.

Farm animals seem to have preferences for particular varieties of seaweed, although the reports of what these preferences are lack consistency. Each of the well-known varieties of brown seaweed growing round our coasts—the stuff known generally as bladder wrack—as well as other varieties such as dulse, honey ware, Irish moss and the Laminarias, have at times been reported as favourite food for farm animals.

Dr. Chapman quotes a dramatic example of the value of seaweed as animal feedingstuff from the *New Zealand Weekly News* of 16th July 1941. 'Some years ago, when I came to this country,' a correspondent wrote in this journal, 'I bought a farm of eight acres. It was the poorest gum land, and in my ignorance I did not realize what that meant. I did not learn that only one cow had been kept there, and I put three on. They died. A veterinary surgeon said that both food and water lacked minerals. I obtained three bags of fish manure and three bags of superphosphate, all I could afford, and then thought of seaweed. It was difficult to get, only the *Hormosira*, or grape variety, being available, and that is the poorest of all, but I managed later to get a little kelp. I soaked the salt out, mixed the chopped weed with a little bran, sometimes a little molasses, and it was wonderful to see the cows go for this food.

'I now have an old cow decidedly growing younger after four years of seaweed rations; another cow, a picture of health, fed on this ration from birth, one little black Jersey, bought when in a starved condition and in calf (she comes in now with four-and-a-half gallons) and two heifers used to the seaweed ration from birth. Nothing has been done to the land all this time, no manure, no lime and no cultivation, and yet the animals are all good and enjoy perfect health. When spring growth of grass comes, one would expect the cows to leave the seaweed ration for the green growth, but not so. They are just as eager for the seaweed then as in winter.'

Sheep in Iceland always feed on seaweed alone for six to eight weeks of the year, although in some years the period may be extended to eighteen weeks. But an even more outstanding example of the value of natural seaweed as animal fodder comes from North Ronaldshay, the most northerly of the Orkney Islands off the north coast of Scotland. This island has its own breed of native sheep, the North Ronaldshay, one of the smallest breeds in the British Isles. The animals take from three to four years to reach maturity, and even then the carcase weight of a wether is only some 30 lb. Their wool is of excellent quality and said to be warmer than any—although fleeces at shearing weigh only 2 lb. or 3 lb., and of this half is usually wasted through contamination. For these reasons attempts have been made by the islanders to cross the pure North Ronaldshay breed with Cheviot, Blackface, Border Leicester and Shetland rams, but without success. None of these breeds could stand the hard and exposed conditions in which the native sheep are kept on the island.

Something like a quarter of the island area is made up of foreshore. It is necessary to cultivate what is left as intensively as possible, and where every available acre is ploughed, there is no room for ordinary sheep farming. This means that the 2,000 or so sheep on North Ronaldshay are restricted to the barren foreshore for the greater part of the year—and thus to a diet composed entirely of seaweed. This is where the local breed scores, for although it is small and less productive than other breeds, it is far hardier. It can live on seaweed. The sheep are purposely restricted to the shore by a stone wall between 6 ft. and 8 ft. high. This runs round the whole of the inner edge of the foreshore, and prevents the sheep straying inland and damaging the crops.

According to D. E. Tribe of the Rowett Research Institute, and E. M. Tribe of the North of Scotland College of Agriculture, who described the North Ronaldshay flock in the autumn 1949 issue of *Scottish Agriculture*, there are some strips of permanent pasture between the wall and the high tide mark at a few points on the coast. But they add: 'It is quite fair to say that the quantity of grass available from these sources could not appreciably affect the animals' diet. There was literally no grass available when we were on the island, and we were assured that even in spring the amount is negligible. The only time that sheep patronize these patches is when an exceptionally high tide submerges other more popular parts of the shore. It is not exaggerating to say that all the sheep except the milking ewes are dependent on seaware for all their nutritional requirements.'

And seaware, according to these observers, is present there in quantity. 'Around the twelve-mile perimeter of the island, where the shore is buffeted on the west by the Atlantic Ocean, and on the east by the North Sea, there accumulate annually hundreds of tons of seaweed; and it is here, on this plentiful supply of food, the supply of which costs nothing, that the sheep live.'

There are stone shelters for the sheep at intervals on the seaward side of the island's ring wall; and immediately after lambing in the third or fourth week of April, ewes and their lambs are collected from the foreshore and folded, or tethered, on grass near the farmhouses for about three months. (This explains the reference earlier to the fact that all the sheep, except the milking ewes, are dependent on seaweed for all their food.) By the beginning of August, however, all ewes and lambs have returned to the foreshore. Any milking ewes and lambs overlooked at folding time usually survive: they are brought in because it is thought wise to let the ewes have good grass at least during suckling.

The North Ronaldshay sheep are never flushed, and only given concentrates if brought in suffering from exposure. They suffer from few diseases. Most of the deaths are caused by drowning or exposure. Sheep scab is unknown. Braxy is uncommon, and there are only isolated examples of gid (sheep staggers). Ticks are entirely absent, and there are exceptionally few worms. Footrot is never troublesome, and bloat unknown.

Wild animals also eat seaweed, including polar bears, rabbits, arctic foxes and wild deer. I have been told by a friend of two deer forests in Morayshire, both much the same as far as grazing and shelter are concerned, although one has access to the sea shore and one has not. Stags shot on the forest with access to the beaches, he tells me, are always a half stone heavier than those shot on the other. The owner and his stalker both believe it is seaweed in the diet which makes all the difference.

Traditionally, seaweed is usually offered farm animals in its natural state, although carrageen stew has been fed to pigs and calves in Ireland, a mixture of boiled *Pelvetia* and oatmeal to calves in the Orkneys, and boiled *Ascophyllum mackaii* to cows in Skye. In the Orkneys, *Fucus* used to be boiled, and the brew poured on hay, chaff, or oat husks as winter feed. In Iceland *Laminaria*, *Fucus* and *Alaria* used to be stored between layers of hay, and used as silage. I have suggested that these practices are of the past, although it may be that some of them are still carried out. I do know that in the nineteenth century and before, cows entered for shows in the Highlands of Scotland would be given seaweed to increase the butterfat content of their milk, and the practice, in more sophisticated fashion, continues to this day. One foodstuff, which is certainly still given in Iceland, is *Ascophyllum nodosum*, dried with heat from hot springs, and used as supplementary cattle fodder.

If seaweed is to be fed to stock in its natural state, it must be near the sea. Seaweed contains much water, and is heavy to carry. That is why the seaweed now used to feed animals is dried and ground near the coast. The resulting meal can be carried at much less cost than seaweed itself, and keeps fresh longer.

In terms of total use, while some seaweed meal is given directly to animals, or mixed with their rations on the farm, the larger proportion is taken by compounders and mixed with other feedingstuffs. This provides balanced rations for dairy cows, pigs, poultry, bloodstock, mink and other fur-bearing animals, as well as household pets.

Most of the seaweed meal imported into this country is used, not on the soil, but to feed animals. As a feedingstuff seaweed is low in carbohydrates, which give energy, and in protein. Apart from calcium and phosphorus, it is rich in trace elements, which are of particular value. It contains fucoxanthin, useful for providing pigment in egg yolks; and contains reasonably high amounts of B and D vitamins.

Here is a table which shows its vitamin content compared with that of grass. It shows that *Ascophyllum* meal (the meal available commercially as 'seaweed meal') provides on a weight-for-weight basis much the same amounts of vitamin A, $B_2$, K and perhaps $D_3$ as dried grass; more $B_{12}$, C and E vitamins, but less niacin.

TABLE 29

Vitamins in seaweed and grass

| | *Ascophyllum meal* | *Laminaria meal (frond)* | *Dried grass* |
|---|---|---|---|
| β-carotene | 16–25 | 11–14 | 100–450 |
| Fucoxanthin (possible precursor of vitamin A) | 90–238 | 469 | — |
| Thiamine ($B_1$) | 1·4–5·4 | 1·3–7·2 | 2·0–3·3 |
| Riboflavin ($B_2$) | 7·5 | 2·4 | 8·6–12·0 |
| Pantothenic acid | 0·2 | 0·28 | 8·6–15·5 |
| Niacin | 12·3 | 19·4 | 29·7–40·7 |
| $B_{12}$ | 0·004–0·08 | 0·06–0·12 | Nil |
| C | 2,674 | 2,094 | Destroyed |
| $D_3$ | 0·01 | n.d. | 0·01–0·03 |
| E | 156–298 | 25·4–29·9 | 111 |
| K | 14·2 | n.d. | 16·7 |
| Folic acid | 0·07 | 0·31 | n.d. |
| Folinic acid | 0·07 | 0·60 | n.d. |

Vitamin contents are given in parts per million. The figures given under the heading *Ascophyllum* meal for thiamine ($B_1$) and pantothenic acid were in fact obtained from an analysis of meal made from *Fucus vesiculosus*, those under the heading *Laminaria* meal for the same vitamins, from an analysis of meal made from *Laminaria saccharina*; n.d. means not determined.

The presence of the anti-sterility vitamin E, or tocopherol, which appears in amounts varying from 1–35 mg. for each 100 gm. of dry matter in seaweed, is worth mentioning in view of the value of seaweed for breeding animals. Eggs deficient in it cannot be hatched successfully, and abortion in cows, and barrenness in sows are said to have been prevented by its use.

Fresh seaweed has one drawback, that its composition can vary seasonally almost as much as grass. This is, however, of less importance in seaweed meal, which is made from seaweed gathered at the same season each year.

Of all these characteristics of seaweed, richness in trace elements—organic minerals which are both safe and potent—is the most valuable. The existence of trace elements is well understood by dairy farmers, who know that lush and otherwise nutritious grass containing adequate amounts of nitrogen, phosphorus and potassium may still fail to supply the needs of grazing stock because certain trace elements are missing, or are not available to the animal. In theory such deficiencies can be corrected by spraying the pasture with a soluble salt containing the necessary element. But diagnosis of a deficiency is not always easy; and even when correct, if too much of the missing element is supplied, the balance may be further upset. Pasture plants and grazing animals both have delicately balanced needs for trace elements; too little may produce deficiency symptoms, too much may be poisonous.

The problem of deficiency in pasture grass is made more severe by modern high-yielding strains of herbage plants, and concentrated artificial fertilizers which encourage the soil to produce crops beyond its natural capacity. These high yields put a heavy strain on the mineral resources of the soil. And even when trace element deficiencies do not lead to obvious disease in grazing animals, minor deficiencies may be responsible for infertility and low resistance to infection.

In the case of many trace elements, nutritional levels and toxic levels are close together. Even if a manufacturer wants to produce a feedingstuff containing the correct amount of trace elements, he cannot be certain that expert opinion, not always infallible, is not exaggerating the amount which can be given without ill effect. Even if the amounts have been correctly assessed, inattention, or difficulties during manufacture, may lead to unsatisfactory mixing, in which case toxic concentrations may appear in certain batches. Even if these difficulties do not arise, feedingstuff with a concentration which is high but harmless for one animal, may be given by mistake to another, with harmful results.

It is true that trace element excess can occur naturally, as in the teart pastures of Somerset, where the grass contains about 300 parts per million of molybdenum instead of the usual 0·2–5·0. This excess molybdenum causes cattle to scour badly, and prevents them absorbing all the copper they need. But in general, and certainly in recent years, if animals get too many mineral elements, it is usually as a result of action by man.

Copper, as a trace element, is valuable to animals, and helps them to make the best possible use of the food they eat. If they get too little they will gradually lose condition, and show such symptoms as stilted gait or diarrhoea. On the other hand, if they get too much over a period it will begin to accumulate in their livers, and will ultimately kill them. Men as well as animals can be affected in this way, but as reports from all over the world of livestock deaths from copper poisoning have shown in recent years, sheep are the chief sufferers. The Edinburgh and East of Scotland College of Agriculture, for example, reported in 1958 that several outbreaks of copper poisoning had been investigated during the year. In one of these, a number of valuable rams were lost. The trouble here arose through feeding a mineral mixture containing copper which had been sold purposely for feeding sheep. Another arose from the feeding of concentrates to which minerals containing copper had been added. The concentration of copper already in the feeding nuts varied between 40 and 60 parts a million, and it had been fed to the sheep for eight months.

'The danger of feeding copper to sheep,' the report of the College continued, 'or allowing them access to minerals containing copper, does not seem to be appreciated by many people marketing such compounds, in spite of the many references in the literature to this hazard'.

The College's annual report for 1959 gave further cases of copper poisoning resulting from feeding mineral mixtures to stock. In one outbreak investigated, deaths occurred six months after the mineral supplement had been stopped. The loss through copper poisoning of thirty-seven in-lamb ewes out of a flock of seventy-eight on a Suffolk farm was described in the *Veterinary Journal* (9th January 1960). The flock was fed a farm-compounded meal rich in copper, which was intended primarily for sixty sows on the same farm. One ton of this ration, which contained bran, maize, fish meal, groundnut meal, wheat, middlings and barley, also contained 60 lb. of minerals with a copper content of 1·75 per cent—enough, as the rations were fed to the ewes, to give them up to 1 gm. each of copper daily. Each ewe was given ½ lb. daily of the compounded meal from 1st to 25th December 1958; then 1 lb. daily for about a fortnight; and thereafter to appetite —up to about 3 lb. daily—until 4th March, when the ration was discontinued.

The first illness was reported on 31st January; antibiotics were ineffective, and on 2nd February two ewes were found dead. A post-mortem showed generalized jaundice of all organs and other symptoms. Copper poisoning was suspected, and confirmed when liver samples showed copper contents, on a dry weight basis, of 1,100, 1,000, 2,100 and 1,600 parts per million. By then, however, it was too late to prevent the death of nearly half the flock.

Cobalt is another metal which animals need, but at the rate of about 1 gm. to 7 tons of dry herbage. More than this increases blood sugar to dangerous levels, and has a harmful effect on the red blood cells. Yet cobalt also has been added to animal feedingstuffs on a scale vastly in excess of the 0·14 parts per million which is the amount found in ordinary pasture grass; indeed, a 7 lb. dressing of one cobalt salt commercially available would give 3,000 times the amount needed.

The fact that excess cobalt is dangerous, both in itself and as a barrier to the absorbing of other trace elements, is one reason why the United States Food and Drug Administration has set upper limits for the cobalt and copper levels in animal feedingstuffs. In the case of cobalt, it is 1 part per million for non-ruminants such as pigs and poultry; in the case of copper, 10 parts per million for sheep. Similar steps have yet to be taken by the authorities in the United Kingdom.

When it was first discovered that parakeratosis in animals was caused by zinc deficiency, and that cattle were suffering from this disease even when eating grass containing 36 parts per million of zinc, it became the practice to add inorganic zinc to animal feedingstuffs. Recommended amounts varied from 50 parts per million advocated by some authorities in Sweden, to 80 or 100 parts per million in the United States, and 200 in Australia. It was then found, however, that pigs fed added zinc sometimes continued to suffer from parakeratosis —it was not realized that the disease was caused, not by a shortage of zinc in the diet, but by an excess of other elements which made the zinc unavailable.

We have already seen that in the case of some elements there can be a narrow margin between the amount needed to prevent disease, and the amount which is poisonous. This latest example shows that there is also a delicate balance between the amounts of the different elements needed, so that excess of one can make another unavailable, and so produce deficiency symptoms.

Another trace element which is necessary in minute amounts, but poisonous if given to excess, is fluorine. One case of fluorine poisoning in a dairy herd, caused by adding commercial mineral supplements to the cows' diet, was described in *The Veterinary Record* of 4th August 1962.

The first sign of disease in the herd of some forty Friesian cows and followers came in June 1961, when lameness became a problem. This lameness was accompanied by loss of condition and reduced milk yield, and it continued in spite of treatment with antibiotics, sulphonamides, corticosteroids, anti-histamines, and other minerals. Urine samples from four of the affected cows showed fluorine levels of 27, 30, 32 and 38 parts per million, much above normal. There was no obvious source of industrial fluorine in the area, and local water had a fluorine content of only 0·4 parts per million. But the herd was given three mineral supplements. One, fed to all the herd, showed low fluorine contents of 515, 600 and 575 when tested. Three samplings of the second mineral supplement, fed to young stock, showed fluorine contents of 2,463, 3,413 and 2,663 parts per million. The third supplement showed fluorine contents of 399 and 453 in two tests. Since this third supplement was fed as 50 per cent of the production ration of the herd it would, when given to the cattle, provide 200 parts per million of fluorine in the ration—as compared with from 60–70 parts per million provided by the supplement with the highest fluorine content of all.

There was then no need to seek further for the cause of lameness in this dairy herd.

The report for 1960 of the Royal Agricultural Society of England had this to say concerning trace elements and their use in feedingstuffs: 'Knowledge of the vital need of farm animals for minerals, and the harmful effect of deficiencies, has steadily increased. The use the body makes of these elements, which may be required in very small amounts, is much more complex than was originally suspected. It is now appreciated that many trace elements are intimately concerned with the function of the enzyme systems: examples are copper, iodine, cobalt, molybdenum, selenium and others, and they are required in very small amounts. Considerably larger quantities of other minerals such as calcium and phosphorus are needed, and the diseases which arise when they are insufficient or absent are well known. Nevertheless there is already evidence that the pendulum may be swinging too far in the other direction. It is not always clearly recognized that an excess can be as harmful as a paucity, and an imbalance is as much to be avoided as an absolute deficiency.

'The administration of super-abundant amounts may arise either through over-enthusiasm on the part of an owner or his staff, particularly when foodstuffs are supplemented which have already had minerals added to them by the food compounder. One example which comes to mind is the case of copper. Hypocuprosis is known in many parts of the world, and was first reported in the United Kingdom in 1946. If the copper intake is low, clinical or sub-clinical disease may arise as evidenced by gradual loss in condition, a stilted gait, diarrhoea, etc. It does not follow, however, that copper is necessarily lacking in the pasture, but that some factor interferes with the absorption of the metal. For example, in the teart areas of Somerset there is an excess of molybdenum, and this adversely affects the absorption of copper. In effect, the amount of copper supplement required is very small, and evidence is accumulating that the quantity which can be tolerated is easily exceeded. In consequence, copper poisoning is on the increase: moreover, apart from the cases which are diagnosed, there are probably many more in which the damage is not acute, and passes unrecognized. Clearly, the need to confirm suspected copper—or indeed other mineral—deficiencies by laboratory procedures is most desirable before incorporating trace elements in the foodstuffs.'

I have said a good deal about the dangers of giving trace elements to stock in inorganic form. The cases I have described are certainly few in comparison with the greater number where mineral additives are given with safety. But the climate of opinion in this, as in related fields, tends to favour the inorganic as compared with the organic, and the very real dangers of the inorganic attitude should be clearly seen. Further, the number of cases of sub-clinical disease following slight excess of inorganic trace elements—or the unavailability of other trace elements which it causes—may be much greater.

Now the difficulties I describe cannot arise with seaweed as the vehicle for trace elements. The amounts of trace elements in seaweed are decided by the plant itself; and while some plants are quite capable of absorbing quantities of minerals which make them as poisonous to animals as a diet containing too much inorganic copper, seaweed is not of their number. The amounts of trace elements in seaweed never approach anywhere near danger point. Further, while in theory it is always possible to produce feedingstuffs with all the trace elements required, and in the minute amounts required, by technological control—well, machines can go wrong. With seaweed, minerals are measured out in absurdly small amounts, and spread throughout a vast bulk of potential feedingstuff, and both processes are carried out completely automatically, and with complete reliability. Where concentration or dilution of trace elements is in question, nothing can go wrong. One knows, with complete confidence, that the toxic effect of giving trace elements can always be avoided by using seaweed as the vehicle.

But in this matter of trace elements seaweed has two further advantages. It has all the trace elements: and it has them in organic form. It has been proved beyond doubt that plants suffering from trace element deficiency absorb chelated minerals where they cannot absorb straight inorganic salts. The valuable tonic effects of organic trace elements even to plants in good health have also been proved. It is now accepted that 'organic' metals are also superior to 'inorganic' where the diet of animals and men is concerned—and that much of the beneficial effect of seaweed in feedingstuffs results from the chelated minerals it contains. The fact that low concentrations of 'organic' elements are just as effective as higher concentrations of 'inorganic' elements also means that the animal's needs can be satisfied without the danger of poisoning, or of making other elements unavailable.

One of the trace elements in seaweed needs particular mention, and that is iodine. Seaweed is indeed the world's best source of organic iodine.

Iodine is combined in seaweed with amino-acids to form the precursor of thyroxine, di-iodo-gorgoic acid, the active principle of the thyroid gland, which regulates the whole of the bodily processes. A deficiency of iodine can cause goitre and, in cattle, stillbirths. And since cows need iodine in direct proportion to the amount of butterfat they produce—2 mg. for each gallon of milk containing 4 per cent butterfat—high-yielding cows, even with normal supplies of iodine, can profit by the extra iodine which seaweed contains.

Thus it was noticed in Russia before the war, when milk yield increased in certain herds from 5,000–9,000 lb. a year, that goitre appeared. When yield dropped during the war, goitre disappeared, only to appear again after the war, when yields increased. The goitre disappeared—and butterfat contents increased—after the herds were fed rations containing iodine.

Certain crops contain what are known as goitrogens, substances which make it difficult for animals to absorb iodine. Among these are kale, swedes and brassicas generally. Non-pregnant ewes fed goitrogenous foods may fail to become pregnant as a result of iodine deficiency; pregnant ewes may suffer from reproduction failure, and their lambs from goitre. These results can be avoided by feeding the flock seaweed.

## TABLE 30

### Trace elements in seaweed (1)

| | *Co* | *Ni* | *Mo* | *Fe* | *Pb* | *Sn* | *Zn* | *V* | *Ti* | *Cr* | *Ag* | *Cu* | *Mn* | *Ba* | *Sr* |
|---|---|---|---|---|---|---|---|---|---|---|---|---|---|---|---|
| *Cladophora rupestris* | 16·2 | 20·0 | 2·44 | 4,400 | 38 | <5 | 92 | 24 | 550 | 6·4 | <0·7 | 31 | 1,260 | 40 | 112 |
| *Rhodymenia palmata* | 2·60 | 16·4 | 0·83 | 1,355 | 28 | <5 | 200 | 29 | 100 | 34 | 1·0 | 48 | 110 | 21 | 90 |
| *Laminaria cloustoni* frond meal | 0·37 | 1·28 | 0·25 | 437 | 7·9 | <5 | 170 | 1·0 | 18 | 1·4 | 0·2 | 4·6 | <20 | 28 | 650 |
| *Laminaria cloustoni* stipe meal | 0·48 | 2·88 | 0·29 | 446 | 5·4 | <5 | 59 | 2·5 | 26 | 1·3 | 0·9 | 5 | 47 | 43 | 2,500 |
| *Ascophyllum nodosum* meal | 1·43 | 3·35 | 1·25 | 1,132 | <4 | <5 | 110 | 5·9 | 114 | 2·9 | <0·3 | 61 | 45 | 27 | 560 |
| Mixed pasture herbage | 0·14 | 3·1 | 0·82 | 56 | 1·6 | 0·4 | 56 | 0·06 | 1·4 | 0·1 | — | 4·6 | — | 108 | 37 |

Contents are expressed in parts per million of dry matter. The elements listed are Co, cobalt; Ni, nickel; Mo, molybdenum; Fe, iron; Pb, lead; Sn, tin; Zn, zinc; V, vanadium; Ti, titanium; Cr, chromium; Ag, silver; Cu, copper; Mn, manganese; Ba, barium; Sr, strontium. The sign < means 'less than'.

A table showing the trace element content of seaweed, and of seaweed compared with pasture grass, may be of value here. It shows the trace elements in two types of untreated seaweed, three types of seaweed meal, and pasture grass (see page 145).

Here is another table which makes a more direct comparison between seaweed meal and an unnamed inorganic salt on the market as providers of trace elements. It shows how much of nine different trace elements is found in a ton of seaweed meal. It also shows the maximum amount of these trace elements lost annually from an acre of pasture—and how much would be replaced if the same acre of pasture were treated with 7 lb. of the salt offered commercially as a source of trace elements.

TABLE 31

Trace elements in seaweed (2)

| | *Supplied by Ascophyllum meal in grams a ton* | *Maximum annual loss from soil in grams an acre* | *Supplied by 7 lb. of a salt in grams an acre* |
|---|---|---|---|
| Vanadium | 6 | 0·12 | 340 |
| Cobalt | 1·4 | 0·28 | 790 |
| Iodine | 500 | 1 | 2,430 |
| Molybdenum | 1·25 | 1·64 | 1,740 |
| Copper | 61 | 9·2 | 810 |
| Boron | 167 | 20 | 362 |
| Iron | 1,132 | 112 | 640 |
| Zinc | 110 | 112 | 724 |
| Manganese | 45 | 190 | 785 |

The figures show clearly that the amounts of vanadium, cobalt, iodine, molybdenum, copper and boron needed by herbage and grazing animal are infinitesimally small—so small that to apply an inorganic salt at the recommended rate of 7 lb. to the acre would provide far and away more of each than plant or animal requires.

It is true, of course, that direct comparisons between the annual loss of a trace element, and the replacement offered in one fertilizer treatment, are not possible. The table also loses something because it does not show, and perhaps could not show, the extent to which the trace elements in salt and seaweed would in practice be available. This is a fundamental weakness. Even so, the table does suggest the dangers of toxic overdressing with an inorganic source of trace elements.

## *Seaweed meal as feedingstuff*

There is one final characteristic of seaweed meal as animal feeding-stuff which must be mentioned. In the same way that seaweed helps plants to make better use of their food, so do the benefits of feeding seaweed to animals seem out of all proportion to the apparent food value of the seaweed itself.

This is a matter of experience. It results from the observations of farmers and breeders, rather than of research workers. Nor is it clear why it should be so. Trace elements and growth hormones are probably involved, although the fact that seaweed modifies the intestinal flora, or bacteria, of livestock may be chiefly responsible.

The part played by intestinal flora may be illustrated with reference to the cow.

Cows (it is no secret) eat grass. Grass is made up of starch, and other nutrients, contained in the cell walls of the leaf. These cell walls are made of cellulose. The cow's digestive juices can deal with the starch and other nutrients, but they cannot break down the cellulose envelope in which they are contained. This has to be done by bacteria.

When a cow eats grass it goes straight into the cow's rumen. When the animal lies down to chew the cud, the grass is brought up into the animal's mouth, chewed, mixed with saliva, and passed back to another part of the rumen. Here it is attacked by intestinal bacteria which break down the cellulose and, in breaking it down, nourish themselves and release starch to nourish the cow. No cow, unless it relied entirely on non-cellulose concentrates such as sugar, could be nourished without the bacteria in its stomach.

The same applies to other animals. No animal produces an enzyme which digests cellulose, yet many feed exclusively on vegetation whose nourishing part is enclosed in cell walls made of cellulose. These walls are partly broken open by chewing, and broken down completely by enzymes produced by the bacteria in the gut of the herbivorous animal—in the caecum and appendix of the rabbit, in the rumen or paunch of sheep and cows.

One cow will produce more and better milk, and thrive better than another cow on exactly the same pasture, simply because of its stomach bacteria. And it is probably because seaweed ministers to the needs of these bacteria that it also improves the health and milk yield of the cow itself.

Other animals, such as hens, pigs—and men—rely on their digestive juices rather than on bacteria for breaking down their food. Nonetheless they also are host to vast numbers of bacteria which help break down such food as is not dealt with by their digestive juices. In the case of men and other animals, these bacteria also help to synthesize vitamin K, and the B complex of vitamins. Races which eat seaweed, as the Japanese do, have presumably conditioned their intestinal bacteria to carry out these duties. It is certainly a matter of common observation that daily doses of seaweed fed to livestock for the first time are not completely digested until after a week or so. It may be that in modifying their intestinal bacteria to help them digest seaweed in this way, livestock are better able to exploit the other ingredients in their diet.

The most spectacular results with seaweed, however, are obtained when it is added to breeding rations. There is no need to stress the importance of any improvement in this direction. If cows do not come on heat regularly; if sows cannot produce enough milk for a litter of eight; if a flock of sheep produces an average of one lamb to each ewe, instead of one and a half; if hens lay eggs which are only 60 per cent hatchable—then success in livestock husbandry can only be partial.

There can be no doubt that seaweed added to the diet does make improvement in these directions. The evidence of practical breeders over many years cannot be gainsaid. I would accept this evidence even if no research had taken place into the effect of seaweed on animal fertility. As it happens, the only fertility trials of which I know have confirmed the value of seaweed in one particular, and I mention them here. These trials were conducted by the Institute of Animal Husbandry at the University of Giessen in West Germany, and concerned the effect on bulls' semen of the addition of seaweed meal to the diet.

Professor I. Kruger's report of this trial is thorough and detailed, and not easily abstracted. It did point out, however, that while extra seaweed had virtually no effect when bulls were fed green fodder, the addition of 400 gm. of seaweed meal a day to the diet of each bull, at times when no green fodder was available, resulted in a 'very high' increase in the number of live sperm, and 'much greater' durability of the semen.

We do not know why seaweed should improve breeding, unless it is the presence of the anti-sterility vitamin E. It may also be that organic iodine and its associated trace elements stimulate the reproductive organs by way of the thyroid gland—but this is only guesswork. I can only say that poor conception rates in cows, the production of abnormal calves, the retention of cleansings, as well as thyroid troubles which result in staring eyes and loss of hair, have all been cleared up by adding seaweed meal to the diet. But not, I hasten to add, overnight. This is an important point. Seaweed meal does not give quick returns when fed to livestock, or human beings, or plants. Slow and scarcely noticeable deterioration in the condition of animal or herd may take months: it is not reasonable to expect improvement to take place in any shorter time.

Here, I think, the modern attitude to medication may be at fault. Most people expect, when run down, to be given medicine which will act quickly. If they are refused drugs, but told to eat certain foods and to allow nature to take its course in other ways, they will usually be disappointed. And doctors, whether they like it or not, are often bulldozed into accepting, and so confirming, this attitude. The pharmaceutical industry needs no encouragement in doing the same. The result is that there is a general tendency for the sick to rely on powerful synthetic drugs in the hope they will produce immediate and lasting results, rather than on more natural long-term remedies based on careful attention to diet, and other simple rules of health.

The patient's motives in seeking quick relief from medicine are mixed: part is natural impatience, part is modern black magic. Further, the philosophy of the age, which rates the synthetic and the mechanical above the natural and organic, is seen at its best in the pharmaceutical industry, where herbal and other remedies are synthesized, and thus produced with less labour and less cost than are involved in growing, cultivating, harvesting, cutting, drying, and otherwise treating natural medicaments. Power, too, can be used in the production of synthetic remedies; with natural remedies, plants must grow, and the mechanization of their husbandry, and their processing, may not be possible. The natural desire of the patient to jump mechanically into good health rather than to grow into it naturally is thus reinforced by what has been called the mechanomorphic philosophy, or superstition, of the age—and the need for money to circulate within the chemical industry.

I have digressed with reluctance. It is not my purpose to justify the use of seaweed on any basis other than that of results; and with plants, whose habits of natural growth are accepted without question, fashionable attitudes of mind are not likely to deny objectivity to observers. But those who accept that plants must grow gradually, and overcome their ailments gradually, often believe men and animals are subject to different laws, and that it is possible for changes to take place in their condition almost overnight. Many sharing this attitude of mind have given seaweed to their flocks or herds; have expected quick results; have been disappointed; and rejected the whole idea out of hand.

It is only because this attitude may automatically prejudice the livestock farmer against treatment which takes weeks, or even months, to take effect, that I have questioned here some of the assumptions on which it may be based.

I must now venture some remarks on the qualitative, rather than on the quantitative, value of seaweed as animal feedingstuff. Increased fertility and improved yield of livestock can be measured, while increased brightness of coats in cows and horses, cleanness of skin in pigs, liveliness in dogs, and docility in mink, are less easily measured and proved. Yet all have been reported to me by those who feed seaweed meal to their stock.

We have seen that it is no use giving seaweed meal to animals to give them energy; and while seaweed can help animals to put on weight, it does this not by offering calories or proteins, but by giving vitamins, trace elements, and other natural products which help them to exploit the proteins and calories in the rest of their diet. Let us see how these, and the other characteristics of seaweed, affect its value as a feedingstuff for individual classes of livestock.

### *Poultry*

Seaweed meal is of more value to breeding and laying fowls than to broilers. The vitamins and minerals help the laying fowl; and seaweed, through the fucoxanthin it contains, improves yolk colour. But since seaweed is low in energy-creating factors, it is of little value to broilers, whose main needs are for energy-producing food and protein in small weight.

Large-scale poultry trials with seaweed as part of the ration took place at the Institute of Poultry and Fur Animals of the Agricultural College of Norway between 1951 and 1954, and again between 1955 and 1959. They were reported, in Norwegian with an English summary, in the 8th and 12th reports of the Institute respectively. Both reports are detailed, illustrated with a number of tables, and contain full bibliographies.

The first series of trials was conducted by the Director of the Institute, Professor Johs. Høie, and Øystein Sandvik. There were two trials with pullets, and six with chicks. I shall deal first with the pullet trials.

*Pullet trials*

A total of 544 White Leghorns were concerned in these trials, the first of which lasted for about seven and a half months during the period 1952–3, the second some eight and a half months in 1954. The hens were given dry mash to appetite, and 40 gm. of corn a day. Hens also received from 25–50 gm. of sour skim milk each, daily, towards the end of the trial period.

The following types of seaweed meal were tested:

(1) *Ascophyllum nodosum*, collected on four occasions between February and July. Part of the salts contained in this seaweed meal had been removed.

(2) *Alaria esculenta*, cut in May.

(3) *Laminaria hyperborea*, *Laminaria saccharina* and *Alaria esculenta* in equal proportions, all harvested in June.

(4) *Laminaria hyperborea*, harvested in May and September.

During the early part of the experiment, when diet was 'somewhat deficient' and egg-laying dropped to between 20 and 30 per cent, egg production improved when 8 per cent of *Ascophyllum* meal (1), autumn harvested *Laminaria* (4), or grass meal, was added to the diet. But when skim milk was later given to all groups of hens, the difference in egg production between seaweed and non-seaweed groups was reduced.

Eight per cent of *Ascophyllum* meal (1) had no effect, and 8 per cent *Laminaria* meal (3) little effect, on hatchability of eggs produced.

In the second experiment, basic diet was 'somewhat superior' to that in the first. Egg production, with no additions to the diet, averaged between 70 and 90 per cent. In these conditions there was no improvement in egg production with the addition, either of 7 per cent *Ascophyllum* meal (1), *Laminaria* meal (3), or grass meal, or of skim milk at the rate of 50 gm. a day. Hatchability of eggs from birds given 7 per cent of either *Ascophyllum* meal (1), *Laminaria* meal (4), or grass meal, averaged 76–86 per cent, compared with 68 per cent in the case of birds on basic rations only.

Professor Høie and Øystein Sandvik, referring to results from both pullet trials, added that no differences in livability could be traced in chicks hatched from eggs laid by the different groups; and *Ascophyllum* and *Laminaria* meal did not colour the egg yolks as much as 8 per cent of grass meal. Iodine content of the yolks of the 7 per cent *Ascophyllum* eggs was 1·46 mg. in each 100 gm. of yolk, in the *Laminaria* eggs 4·55 mg. Iodine content of non-seaweed egg yolks was

0·034 mg. The addition of *Ascophyllum* or *Laminaria* meal had no effect on storability, smell, or taste of the eggs.

*Chick trials*

In the chick trials a total of 2,356 chicks were involved. They were kept from one day old to eight weeks, were housed on wire-floored battery cages, and fed dry mash and water. The same types of seaweed meal were used as in the pullet trials.

When 3–4 per cent *Ascophyllum* meal (1), or 4 per cent *Alaria* meal (2) was added to a well-balanced chick feed to which yeast, cod liver oil and, in one instance, grass meal, had been added, the results were no better than with the basic diet alone. But when *Ascophyllum* meal (1) and *Laminaria* meal (3) and (4) were added to a well-balanced chick feed, but one without yeast, grass meal, or cod liver oil, the seaweed meals produced in some cases a great additional gain in weight, and an improvement in health.

Some of the most dramatic results, however, were obtained when meal made from young fronds of May-harvested *Laminaria hyperborea*, one of the two types mentioned under (4), was used to make up 5 per cent of a chick mash without yeast, grass meal, or cod liver oil. Chicks fed this ration in the first six weeks of their life increased in weight 30 per cent more than chicks given the same ration without seaweed. In the first eight weeks of their life the percentage weight increase was 22 per cent. But when the 5 per cent addition was made up of *Laminaria hyperborea* harvested in September (4), then the additional weight increases were only 9·8 per cent and 3·5 per cent respectively.

The addition of yeast to the chick feed, however, resulted in an even greater gain in weight than the addition of seaweed meal; and chicks given seaweed meal and yeast gained more weight than those given seaweed meal alone.

I will quote the authors' comments on these chick trials in full. 'A major reason for the (weight increase) effects obtained by feeding seaweed meal, is that these stuffs probably increased the content of riboflavin in the feed. On the other hand it is difficult, on the basis of our experiments, to decide whether the seaweed meal had any vitamin A or D effect. The addition of cod liver oil to the basic diet alone had in fact no effect on the gain in weight, or the health condition. Accordingly, the basic diet had to contain enough vitamin A and D to meet the requirement involved in the relatively slight gain in weight produced by the basic diet alone. In one experiment in which we used chicks with small reserves of vitamins from the eggs, and where we had obtained pronounced effects by adding seaweed meal, or grass meal, an addition of cod liver oil produced great effect when fed together with yeast or grass meal, but none when given together with seaweed meal alone. The increase in growth resulting from the addition of yeast must have increased the requirement for vitamin A and/or D, so that the quantity contained in the basic diet, plus that possibly added through the seven per cent of the *Ascophyllum* meal, was not sufficient for these chicks.'

Commenting on the 1951–4 trials on both chicks and pullets, the two authors pointed out that the addition of *Ascophyllum* or *Laminaria* meal to chick or hen rations produced no noticeable results when those rations were well balanced. But if the meals were added to a chick feed lacking yeast or grass meal, but otherwise well balanced, then growth was more vigorous, and health was improved. The results then obtained were much the same as those which resulted from the addition of grass meal to the same ration.

Additions of seaweed meal to hen rations without yeast or grass meal, but otherwise well balanced, had no noticeable effect on egg-laying; but when hens received no milk, hatching results were improved—and egg-laying as well, if the hens were not kept on litter.

Addition of seaweed meal had no noticeable effect on storability, flavour, or odour of the eggs; effect on yolk colour was slight, although additions of *Ascophyllum*, and particularly *Laminaria*, meal resulted in a great increase in the iodine content of the eggs.

The authors also pointed out that the fact that spring-harvested *Laminaria* produced better results with chicks than autumn-harvested *Laminaria* was evidence that the value of seaweed, like that of other plants, might vary with plant species, and time of harvest.

*Chick and hen trials*

The second main group of trials, between 1955 and 1959, were conducted by Professor Høie and Fridtjov Sannan on 138 laying hens and 3,052 chicks. Both chicks and hens were housed in battery cages on wire, and given experimental feeds and water only. The chick trials lasted from day-old to six or eight weeks; the hen trials from six weeks to eight months. Chicks were given *Ascophyllum nodosum*, *Fucus vesiculosus*, *Fucus serratus* and *Laminaria hyperborea* meal; laying birds were given *Ascophyllum nodosum*, *Fucus vesiculosus* and *Laminaria digitata* meal.

In the chick trials, when from 5–7 per cent of *Ascophyllum*, or 5 per cent of the other three types of meal listed, was added to all-round basic diets which included yeast and vitamins A, D and $B_2$, the growth rate, feed efficiency and health of the chicks were unaffected. When these meals were added at the rate of from 10–15 per cent of the ration, and energy and protein content were not made up accordingly, the chicks' health was unaffected, but their growth, feed consumption and feed efficiency suffered.

On the other hand, when the basic diet was low in vitamins, but otherwise well balanced, there was a clear response in growth rate, feed efficiency and health when 5–7 per cent of seaweed meal, or of grass meal, was added to the ration. When the grass meal was of good quality, it produced responses equal to that of added seaweed; when of poor quality, responses were not so good. Slight differences in the effect on growth, and in the reduction of vitamin deficiency, of the four types of seaweed meal were recorded.

The addition of 2 or 3 per cent of yeast had the same effect on growth as the addition of seaweed meal in two of the experiments, although in a third experiment it produced better results.

The best growth rate, feeding efficiency and health were obtained when vitamin supplements were added to a ration low in vitamins, either alone or with yeast. This suggested that neither additional grass meal nor seaweed meal was enough as sole supplement to a basic ration low in vitamins.

In one experiment, the addition of from 5–7 per cent of *Fucus vesiculosus* and *Fucus serratus* meal had a laxative effect on chicks aged two or three weeks. This effect wore off later. When, in two other experiments, from 10–15 per cent of different seaweed meals were given, it had no unfavourable effect on the chicks' digestion.

It was found in experiments with a basic diet low in vitamins that seaweed meals (as well as grass meal) had a pronounced vitamin A effect. There was a 'relatively pronounced' riboflavin effect with seaweed meals (and grass meal) in two experiments, and a slight riboflavin effect with *Ascophyllum* in three further experiments. In two other experiments, where such effect might be expected, none was observed. It was thought that this might have been caused by variations in the riboflavin content of the meals.

The authors further commented on these chick trials: 'Relatively small variations in the vitamin B content of the basic diets may be responsible for varying effects when seaweed meal is added to the diet.

'Addition of five per cent grass meal, or *Fucus* seaweed meals containing no special sources of vitamin E, as a rule produces a considerable increase in vitamin E content. Thus part of the responses obtained by supplementing the rations with grass meal and seaweed meals may be caused by vitamin E.

'The layout of the experiments made it impossible to obtain satisfactory information on the value of the minerals in seaweed meals for chick feeding. However, in a number of experiments the mineral supplementation of the basic diets consisted of ground limestone, di-calcium phosphate, and sodium chloride (common salt) only, with no special sources of trace minerals. It seems therefore possible that the content of trace minerals in seaweed meal may have influenced responses obtained by supplementing the diets with seaweed meals.'

A further trial was that to discover how much seaweed meal hens would tolerate. Five per cent of *Ascophyllum*, *Fucus vesiculosus* or *Laminaria digitata* in an all-mash diet had no effect on droppings, although when the proportions were increased to 10 per cent, and particularly 15 per cent, wet droppings resulted. These results were the same whatever the date of harvesting. The authors commented that the addition of from 5–7 per cent seaweed meal to an all-mash meal would cause no digestive disturbances in hens.

In another trial there was no drop in egg production even after 15 per cent seaweed meal had been given to laying birds for three months. The diet was all mash, and was not corrected for the drop in energy and protein content. (Annual egg production on the basic diet was not given in the English summary of the report.)

When 5 per cent seaweed meal was added to the diet it had a 'positive effect' on the colour of egg yolks, and this effect became more pronounced as the percentage of added seaweed increased. In a few cases the addition of from 10–15 per cent of specially ground seaweed meal gave a reddish colour to the yolks, an effect which was not produced by the same amounts of commercially produced seaweed meals. The authors commented: 'It seems obvious that meals of *Ascophyllum nodosum*, *Fucus vesiculosus*, and *Laminaria digitata* may contain carotenoids, which by some hens can be conveyed from the feed to the eggs, giving the yolks a reddish colour.'

Apart from colouring the yolks, seaweed meals had no effect on egg quality or storability.

The authors summarized their findings in the second series of trials by saying that supplementation with seaweed meal of chick rations deficient in certain vitamins, particularly A and B, resulted in a marked improvement in growth, feed efficiency and health. They also pointed out that grass meal and feed yeast had comparable results, and added: 'It should be noted that the value of seaweed meal as a supplementary feed may vary widely with the habitat of, and harvesting time for, the seaweeds, with methods employed in the production of the meal, manner and duration of storage, and so on.'

Added seaweed meal produced much the same responses as ordinary commercial grass meal; and the experimental and analytical results showed that one could count on a 'pronounced effect of vitamins A and E, partly also vitamin $B_2$, when grass meal and seaweed meals are used to supplement diets low in vitamins'.

As to growth rate, feed efficiency and health in the chicks, results were decidedly better when vitamin supplements were given. 'In well-balanced diets for chicks and laying hens,' the authors added, 'supplementation with seaweed meals was not shown to have any special, positive effect on production.

'The experiments show that when all-round mixtures are corrected for decrease in energy and protein content resulting from supplementation with seaweed meals, at least seven per cent of the seaweed in an all-mash diet for chicks, and from ten to fifteen per cent for laying hens, can be used without any noticeably unfavourable effects on health and production.'

The authors pointed out in conclusion that they did not set out to investigate the mineral effect of the supplementary feeds, and that trials to this end would probably have produced even more satisfactory results.

It will be noticed that seaweed meal had little effect on yolk colour in the early trials, but had a pronounced effect in the latter trial. This resulted from improved production methods which fully retained the fucoxanthin in the meal. This influence of fucoxanthin on yolk colour led Mr. Jensen to examine the effect in greater detail. He was able to show (in *Poultry Science*, 1963) that carotene had no influence on yolk colour, which was affected only by the xanthophyll, fucoxanthin. He was also able to trace chemical modifications in the fucoxanthin as it passed through the hen's intestinal tract to the egg—a chemical change to a xanthophyll which produced an even stronger colour than fucoxanthin.

So much for poultry trials. I could have wished that trials on such a scale had taken more account of the effect on poultry of the mineral content of seaweed meal. And even though my account of the various trials is a summary of the original report, it will doubtless strike the general reader as complicated. I felt, however, that poultry keepers would not welcome a less detailed summary than I have given here.

Practical experience—or so chicken farmers tell me—suggests that feather picking and cannibalism, already known to be reduced by the addition of low energy–high fibre ingredients to the diet, become less when seaweed is added to poultry rations. Since shell quality is also much influenced by the supply of trace elements, I would expect seaweed to be of benefit in this direction, although I know of no evidence on these lines arising from experiments.

It is, however, accepted that seaweed restores the red colouring lost from exotic birds' plumage as a result of captivity. When the flamingoes in the zoo at Basle lost their pink colouring recently, it was restored by giving them seaweed meal.

*Sheep*

Sheep are usually fed on low-energy diets, and sometimes on land with mineral deficiencies. Since this means that sheep are not adapted to a high-energy diet, and may suffer from trace element deficiencies, seaweed can make a useful addition to their diet.

This fact has been tested on a wide scale by Erik A. Saeter and Arne Jensen of the Norwegian Seaweed Institute, in trials which continued for nearly eight months. They involved 1,800 ewes on sixty-seven highland farms, the following breeds being represented: Dala 60 per cent, Spel 30 per cent and Cheviot 10 per cent.

The sheep were divided into two groups, each of 900 animals, as equal as possible in breed, body weight, age, wool production, number of lambs in previous years, and so on.

Stall-feeding of the two groups began in the autumn of 1954, from which time ewes in the experimental group were given 35 gm. (1¼ oz.) of *Ascophyllum nodosum* meal in addition to their daily ration. This continued until the end of the lambing season in spring of 1955, when the ewes were sent to mountain pastures. Apart from the seaweed given to the experimental group, feeding was identical in experimental and control groups.

The most positive result of giving seaweed meal was to increase winter wool production. In flocks where neither herring meal nor mineral supplements were given, it increased by an average of 20 per cent, mainly as a result of reducing moulting. 'The number of moulters was undoubtedly reduced by addition of seaweed meal to the rations,' Saeter and Jensen report.

A second result was that ewes in the seaweed group lost less weight in winter, particularly in flocks which were given no herring meal or mineral supplement. Lambs born to ewes of the seaweed group also grew faster than those in the control group. As a result, when the time came for lambs from both groups to be sent to the mountains, 'seaweed' lambs from the Dala breed were heavier than those of the same breed in the control group—and were still heavier on their return from the mountains at the end of the season.

Lambs of Spel and Cheviot breed, however, were of roughly equal weight by the time they were sent to the mountains for the first time.

The influence of seaweed on fertility was less obvious, although on a few farms the extra seaweed meal 'led to a significant increase in the number of lambs born per ewe'.

Repeat breeding, sterility, number of intestinal parasites, weight of new-born lambs born to each ewe and, with one exception, mortality of suckling lambs, these were all unaffected either way by the addition of seaweed meal to the diet. Loss of suckling lambs was, however, greater in control than in experimental groups in all areas save Alvdal and Tynset, which had 20 per cent of the sheep under test.

There were more deaths from pulpy kidney disease and dystocia in experimental than in control groups in all districts. (Pulpy kidney is usually commoner among the biggest and best lambs, while dystocia is often a result of bearing a big lamb.) There was also a marked reduction in the loss of animals from white muscle disease among the 'seaweed' animals.

The two authors imply that these results might have been even better had not the summer of 1954 been 'comparatively' rainy. 'Experiments carried out in two successive years', they report, 'strongly indicated that the effects of the seaweed meal addition were dependent on the quality of the hay given. The seaweed meal is expected to give better results when given after a dry summer, than when used after a wet one.'

Little experimental work apart from this Norwegian trial has been done with sheep and seaweed—the practical work done on North Ronaldshay perhaps makes it unnecessary. But a small trial undertaken by workers in the animal nutrition department of University College, Dublin, to find out the value of three types of seaweed meal—*Laminaria*, *Ascophyllum* and *Fucus*—as sources of energy and digestible organic matter, produced two interesting results. They found that *Fucus* contained only small amounts of digestible organic matter (a finding which directly contradicts the results of trials in this country); and that *Laminaria* and *Ascophyllum* contained considerably more digestible nutrients than *Fucus*, samples taken in autumn being 'comparable in feeding value with meadow hay'.

*Cattle*

Seaweed is valuable for dairy cows in low concentrations because it provides trace elements. There is also sound evidence that it improves milk quality, including butterfat content. It should also be valuable in providing minerals and vitamins to cows brought inside on highly concentrated rations, or to stock on land with trace element deficiencies. Breeding beef cows, if fed on anything other than green stuff, should also profit from the addition of seaweed meal to their diet. Seaweed meal cannot, however, be regarded as a completely balanced mineral supplement for dairy cattle because it cannot provide all the phosphorus, and perhaps also all the calcium, which the dairy cow needs.

Here is a table showing minerals and vitamins made available to a cow when 10 per cent of a 10¼ lb. concentrate ration is made up of seaweed. It refers to the needs of dairy cows only, and shows that while poorer than grass in phosphorus and magnesium, seaweed is richer in calcium, common salt and iodine.

TABLE 32

Seaweed for dairy cows

Minerals and vitamins available to dairy cow when 10 per cent of 10¼ lb. daily concentrate ration is made up of seaweed

| | *Cladophora rupestris* | *Rhodymenia palmata* | *Laminaria frond* | *Ascophyllum* | *Maintenance requirements of 1,000 lb. cow* |
|---|---|---|---|---|---|
| Ca | 7·06 | 3·34 | 4·84 | 10·05 | 10 |
| P | 1·25 | 2·60 | 1·30 | 0·42 | 10 |
| NaCl | 29·6 | 24·6 | 34·1 | 34·3 | 28 |
| Mg | 3·39 | 1·81 | 2·70 | 3·81 | 10 |
| I | 0·51 | 0·14 | 2·32 | 0·23 | 0·20 |
| A | — | — | 15,500 | 31,000 | 40,000 |
| D | — | — | — | 460 | 3,000 |

Minerals shown are Ca, calcium; P, phosphorus; NaCl, common salt; Mg, magnesium; I, iodine; with vitamins A and D. Weights of minerals in grams, vitamins in international units.

Trials to test the effect of seaweed on dairy cows have been carried out in this country and Norway. The best known were those conducted in Scotland by Dr. G. Dunlop of the West of Scotland College of Agriculture at Auchincruive in Ayrshire. Dr. Dunlop had already carried out trials to test the effect of inorganic copper—copper sulphate—on the milk production of dairy cows. In these earlier trials he found that butterfat production could be increased an average of 20·1 gm. a day if cows were fed copper sulphate in their rations. He assumed that it was the copper which produced this effect, but he discontinued the trials because he feared that continued doses of an inorganic copper salt might harm the cows' constitution.

In the seaweed trials he persuaded the owners of seventeen dairy herds in the counties of Ayr, Lanark, Wigtown, Dumfries and Renfrew to give chosen cows 7 oz. of seaweed meal a day. Results were noted by the milk recorder. In all, 108 pairs of cows of similar age, which had calved in the same month, and had similar milk yields and butterfat contents, were used in the trials. One cow in each pair was given her ordinary ration, the other her ordinary ration less 7 oz. This loss was made up by seaweed meal.

The cows remained on the farms. This is a method which appeals to farmers, who believe the production of cows may be affected by moving them into strange surroundings. It is suspect to some research workers, who believe that control is difficult, and variables may creep in, where trials are not conducted in laboratory conditions. The reader can draw his own conclusions in the matter. I may add, however, that to conduct feeding trials on the farm is the general practice on the Continent.

High-yielding cows took to the seaweed meal readily. Only cows with a yield of below 3½ gallons a day refused it, but what was left over was usually eaten the next morning. In five of the herds, farmers stopped the experiment because low yielders refused to continue eating seaweed. The twelve herds which remained were fed seaweed for a further three or four weeks.

All the cows which ate seaweed produced milk with a higher butterfat content—in some cases, between 90 gm. and 100 gm. a day. These butterfat increases are recorded in the diagram on page 161. Two herds where the number of cows tested was thought to be too low—five pairs on one farm, and two on the other—are not shown, although there was an increase in butterfat in both cases.

Dr. Dunlop noticed that the inclusion of seaweed in the ration for three weeks stimulated fat production in some herds to a greater extent than a single 10 gm. dose of copper sulphate. He also decided that the effect of minute amounts of copper in seaweed was more effective, and longer-lasting, than large doses of copper sulphate—in a word, that seaweed was a more effective source of copper than an inorganic copper salt as far as animals were concerned. The suggestion is one which has received some confirmation from experience in the Rowett (animal nutrition) Research Institute at Aberdeen, where it has been found that organic copper is three times as 'available' to the body as inorganic copper.

TABLE 33

Dunlop trials

Daily increases in butterfat among dairy herds in south-west Scotland following addition of seaweed meal to diet. Increases in grams.

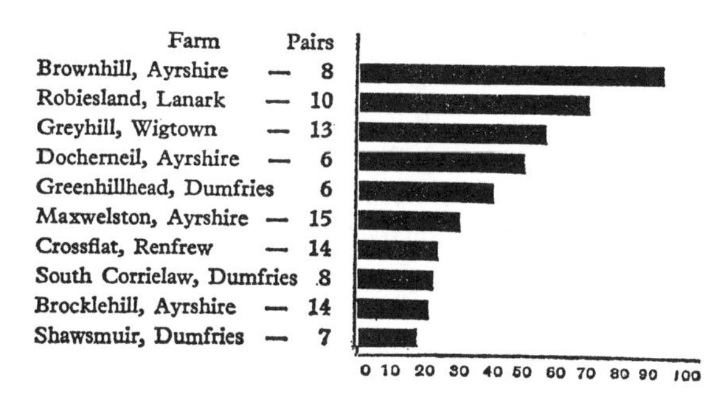


Subsequent trials at the National Institute for Research in Dairying, Reading University, did not confirm Dunlop's findings. In these trials, 10 per cent of the cows' feed was replaced by seaweed meals, or by a mixture of $8\frac{3}{4}$ per cent oatfeed, and $1\frac{1}{4}$ per cent salt. The seaweed material was fed for two periods of three weeks, either separated by a three-week period without seaweed, or for six weeks at a stretch. Those who conducted the trials commented: 'No significant differences in fat production occurred between the treatments, in contrast to the conclusions of Dunlop, in spite of feeding amounts of seaweed meal which averaged 1·3 pounds a cow a day, a quantity considerably greater than was used by him.' No doubt this feeding of so great a quantity explains why no significant differences occurred; moreover, no time was given to the animals to adjust themselves to the change of diet.

The Norwegian cattle trials to which I referred earlier were begun near Trondheim five years ago by Harald Nebb and Arne Jensen of the Norwegian Seaweed Research Institute, and are still in progress at the time of writing.

Identical twin cows are being used. Their basic rations, typical of those given in this part of Norway, are hay, raw potatoes, concentrates, grass and silage. In addition, each control twin is given 100 gm. daily of a standard mineral mixture, and each test twin 200 gm. of a fortified seaweed meal. The mineral mixture contains the following elements: phosphorus 8 per cent, calcium 11, magnesium 15·5, sodium 9, chlorine 14, iron 0·1, cobalt 0·01 and iodine 0·0075. According to the two workers, the mineral content of 200 gm. of seaweed meal 'corresponds well' with that of 100 gm. of the mineral mixture. Apart from these two supplements, the twins are given identical rations.

In all save one of the six sets of twins, the cow fed fortified seaweed has produced more milk than its test twin—on average, some 150 kg. of milk more for each lactation. This amounts to an increase of 4·5 per cent in total milk production. The inclusion of seaweed meal in the ration has not influenced the fat content of the milk—a finding at variance with Dr. Dunlop's conclusions.

These findings are not necessarily in conflict. The effect of adding to the basic ration of ruminants seems to depend on the nature of that basic ration, and on the proportion of fatty acids being produced in the rumen. These in turn influence the amounts of butterfat, total milk, non-fatty solids and body fat produced by the cow.

On a low-roughage ration, yields tend to be high, and butterfat low. The addition of seaweed to such a ration might well increase butterfat. On a high-roughage ration the reverse would probably be true. In any case, it is almost impossible to increase both yield and butterfat at the same time—there is a strong tendency for the two to move inversely, because fatty acids tend to favour (or discourage) milk yield, while discouraging (or favouring) butterfat content; and the matter is decided by the proportions of each type of fatty acid produced.

I end this section on seaweed as feedingstuff for cattle with the report of a further trial conducted by the Norwegian Institute of Seaweed Research. In this trial an increase of between 5 and 6 per cent in milk production, as well as a higher butterfat content, resulted from the substitution of a seaweed–mineral mixture for a standard mineral mixture.

The seaweed mixture contained (about) 79 per cent seaweed meal, 20 per cent di-calcium-phosphate, 1·2 per cent magnesium oxide and 0·06 per cent copper sulphate. The mineral mixture was a standardized one containing magnesium. The cows in the trials were low yielders—the highest yield quoted was 8,434 gallons at 4 per cent butterfat—and it is possible that high-yielding cows might not have responded so well. Nor was the composition of the standard mineral mixture given. But subject to these factors, the trial shows that 200 gm. of fortified seaweed meal a cow a day are more than equal to the 100 gm. of the Norwegian standardized mineral feed usually recommended.

The reader may feel that the result of this trial, in which a seaweed-mineral mixture is proved superior to a pure mineral mixture from the point of view of milk production, runs counter to our main argument, that minerals are best given in organic form. I felt it necessary to report the trial, however, on the principle that all relevant evidence should be given, even that which appears contradictory.

Here, however, we are concerned with something which tends to obscure, rather than contradict. Not only do we not know the content of the standard mineral mixture, but we do not know whether added seaweed meal might not have given the same, or better, results in the complete absence of all inorganic minerals. The results of the trial are therefore at best partial, and further trials will doubtless take the matter further. Even so, the trial is reported here.

*Pigs*

The following table shows the extent to which different vitamins are provided by two types of seaweed meal when fed as 10 per cent of the ration, and how much vitamins porkers, pigs and sows need. The table shows that seaweed provides all the vitamins A and D a pig needs, and partly satisfies its needs of the vitamins $B_1$, $B_2$, $B_{12}$, and nicotinic and pantothenic acids.

Trials in Ireland during the war suggested that seaweed was two-and-a-half times more valuable than potatoes as pig food, and intermediate in value between hay and oats. The trials were conducted by Professor E. J. Sheehy of University College, Dublin, and others, primarily to test the digestibility of seaweed meal. *Laminaria* seaweed was used, dried to 10 per cent moisture, then ground and fed to the pigs with other food.

Professor Sheehy made the point that the seaweed used was autumn harvested and rich in laminarin, and that this laminarin was completely digested by the pigs. As a result of these trials he decided that the digestibility coefficient of the meal was 67 per cent for organic matter and 75 per cent for nitrogen-free extract; and that total digestible nutrients of the meal could be expressed as a factor of 50, compared with 60 for oats, 40 for hay and 19 for potatoes.

Professor Sheehy qualified these findings, however, by saying that his estimate of total digestible nutrients was based on the assumption that his figures for organic matter and nitrogen-free extract were 'true rather than apparent'; and it is necessary to follow his reasoning further on this point.

TABLE 34

Seaweed for pigs

| | *Vitamins provided when 10 per cent of 1 lb. of ration is made up of seaweed* | | *Vitamins needed in each pound of a pig's ration* | | |
|---|---|---|---|---|---|
| | *Ascophyllum meal* | *Laminaria meal* | *Porkers 5–12 weeks* | *Pigs 3–6 months* | *Sows* |
| A | 3,000 i.u. | 1,500 i.u. | 500 i.u. | 600 i.u. | 2,200 i.u. |
| $D_3$ | 46 i.u. | — | 50 i.u. | 50 i.u. | 50 i.u. |
| $B_1$ | 0·15 mg. | 0·19 mg. | 0·5–0·8 mg. | 0·5–0·8 mg. | 0·5 mg. |
| $B_2$ | 0·34 mg. | 0·11 mg. | 0·8 mg. | 0·8 mg. | 0·8 mg. |
| Nicotinic acid | 0·58 mg. | 0·88 mg. | 2·2–5·0 mg. | 2·2–5·0 mg. | 2·2–5·0 mg. |
| Pantothenic acid | <0·01 mg. | 0·013 mg. | 3·4–4·5 mg. | 4·5 mg. | — |
| $B_{12}$ | 0·003 mg. | 0·004 mg. | 0·009 mg. | 0·009 mg. | 0·009 mg. |

i.u. means international units, mg. milligrams, and < less than

He said seaweed meal could absorb a good deal of water. It first formed a thick stiff jelly and then, as more water was absorbed, became a glutinous or slimy mass. Fed with discretion, such material had a favourable influence on the action of the large bowel. It dilated it without irritating it, so helped its movement, and made evacuation regular and complete. He pointed out that foods of this type, when included in a complete diet, not only favoured the gloss of coat and bloom of skin associated with good health, but might also enhance the value of other foods. It was therefore possible that the high digestibility figures obtained in the trials were more apparent than real, in that the seaweed increased the nutritive value of the basic ration with which it was fed.

'But the result is the same,' Professor Sheehy commented, 'whether the extra nutriment absorbed when seaweed is added to a basic ration comes entirely from the seaweed—or partly from it, and partly as a result of improved utilization of the basic ration. That the latter is the more correct explanation is suggested by the fact that hydrolized weed (a liquid solution of seaweed also used in the trials) did not prove superior to the non-hydrolized—the hydrolized weed lost its gelling properties, and presumably its favourable physical effect on the alimentary tract.'

Others have noticed that the addition of seaweed meal to an animal's diet seems to make it possible for the animal to get more out of the rest of its food than would have been possible without added seaweed. They, however, suggest that vitamins, trace elements, or enzymes in the seaweed may exert some sort of catalytic action.

C. D. T. Cameron, writing in the *Canadian Journal of Agricultural Science* for 1954, said that 2–6 per cent of seaweed meal made up largely of *Ascophyllum nodosum* might be added to pig rations with safety, although there would be no increase in growth rate when the meal was added to a well-balanced ration. He added that a mineral supplement was unnecessary when 6 per cent of a pig's ration was made up of seaweed meal.

Harald Nebb and Arne Jensen of the Norwegian Seaweed Research Institute have conducted a test to see whether seaweed meal fortified with calcium, phosphorus and vitamin D could be substituted for the mineral-and-vitamin mixtures used in commercially compounded pig foods. As sometimes happens, the most interesting finding here was unexpected. There were fifteen discarded livers among the control pigs, but among the pigs fed seaweed, only six. As a result of these findings, the possibility that seaweed meal reduces liver parasites in pigs is to be investigated by the Institute.

The other findings in this trial were not of much importance. The seaweed groups had less back fat, and their carcases tended to be longer. There was no difference in daily liveweight gain, food consumption, food conversion ration and carcase quality between pigs given fortified seaweed and those given none, which led the two researchers to comment: 'The conclusions to be drawn from the feeding experiments with bacon pigs must be, that three per cent of seaweed meal, fortified with calcium, phosphorus and vitamin D, can supply all the extra minerals and vitamins needed in bacon production under normal conditions.'

A further report from Norway on seaweed for pigs was published in 1960 by Arne Jensen of the Norwegian Seaweed Research Institute, and Johannes Minsaas. They found that if the 3 per cent of ground barley given fatteners from 44 lb. to bacon weight was replaced by the same percentage of *Ascophyllum nodosum* meal, then the pigs grew better, and made better use of their ration, than a control group. The animals in these trials were given a limited ration designed according to the standard growth curve for pigs. All rations were supplied with the requisite amount of calcium and phosphorus; but no trace elements or vitamins were given, except naturally in the seaweed meal. One would expect the vitamin A, zinc, and other trace elements in the seaweed to offset small reductions in energy and protein; and this is what happened in these trials. 'Pigs on rations with seaweed meal showed better growth and feed conversion than the control group,' the two workers reported, adding: 'No difference in carcase quality was noted.'

Similar results were obtained in a second trial, where 40 per cent of the total energy requirements of fatteners from 44 lb. liveweight until bacon weight was given in the form of boiled potatoes. Ground barley, and a compound feed containing some 30 per cent digestible protein (which contained vitamins, minerals and trace elements) were given in addition to the potatoes, and the whole ration limited according to the standard pig growth curve. 'Pigs on the ration containing seaweed meal', Jensen and Minsaas reported, 'showed slightly better growth and feed conversion than the control.' Their carcases were, however, slightly less firm. Apart from this, there was no difference in carcase quality or killing-out percentage between the two groups, although there were savings of 2–6 per cent of the feed bill as a result of using seaweed meal, which costs less in Norway than feedingstuff.

In nearly all trials with bacon pigs, the addition of up to 5 per cent seaweed meal to rations has had no adverse effect, either on growth, or on the rate of food conversion. In one case, however, a Norwegian researcher, T. Homb, found that the addition of seaweed to a baconer's diet at percentages as low as 3–6 per cent reduced daily liveweight gain.

It must, of course, be remembered that seaweed is a source of vitamins and minerals, rather than of protein and energy-producing food. And in this trial the meal was added to a ration already low in protein, and one already fully supplied with minerals and vitamins. This meant that the seaweed-containing diet fed to baconers in Homb's trials had a lower protein and energy-producing content than the rations fed the control pigs. This much, analysis of the two rations shows. And since it is well known that even small reductions in the protein content of an animal's diet may lower its growth rate—particularly when protein levels in that diet are already low—it is not surprising that the seaweed-fed baconers fared badly. However, as much of the other work quoted demonstrates, if up to 5 per cent of seaweed is included in rations in which protein or energy is not at a critical level, no adverse effect is produced. Indeed, as we have already seen, the special physical properties of seaweed meal may increase the efficiency with which the rest of the ration is used. Further, where breeding stock are concerned, supplies of protein and energy are rarely limiting factors, while the long-term beneficial effects of seaweed meal are especially desirable.

*Horses*

Horses respond well to seaweed in their diet—their delicately balanced mineral needs can be safely satisfied in this way. A number of breeders in this country give their brood mares seaweed to increase fertility, while in the United States it has been found that seaweed in the diet prevents cracked hooves. Lymphangitis, an inflammation of the nerve sheaths for which there was at that time no known cure, was treated successfully in French army horses after the First World War by giving them seaweed to eat.

*Mink*

Some years ago, at a time when mink breeders in the United Kingdom had scarcely even heard of seaweed products, our United States distributor told us that they were the largest class of breeders he supplied. Mink breeders in this country have since begun to give seaweed meal to their charges, and my own company now sells them something like 20 tons of meal a year, in 1–5 cwt. lots. They say that since giving breeding stock ½ oz. of seaweed meal a day they have had the best and strongest kits (litters) they have ever known; that survival rates have improved; and that the mink are more docile.

Mink are notorious as one of the most unpleasant-tempered of all 'domestic' animals. They cannot be tamed. They can never be trusted. Given half a chance, they will bite anyone or anything. But the fact remains that mink breeders, who are at the receiving end of their charges' attentions, say seaweed meal makes them more docile; and I repeat their claims here.

These statements are to some extent confirmed by a recent trial in Finland. Fifty mink at the Ollikkala mink farm school were given seaweed meal in their rations and produced 182 kits, an average of 3·64 young for each female. Fifty mink not given seaweed meal produced 166 kits, an average for each female of 3·32. Two females in the non-seaweed group, however, died during the course of the trial for reasons which were not investigated, so that the average number of young for each female in this group should be a little higher—3·45.

The trial began on 1st January 1965, and ended on 30th April, and mink given seaweed meal came on heat three days earlier than the control group. Both groups were given a proprietary feed, seaweed meal making up 3 per cent of the total ration in the trial group.

*General*

A review of the whole field of seaweed and animal feeding shows that the value of seaweed has been proved by experience rather than research. It is also true that the findings of research, while in general favourable to seaweed, particularly as a provider of trace elements and vitamins, are in some ways conflicting.

There are at least two facts which go some way towards explaining these conflicting findings. The first is that while each seaweed processer tries to standardize his own product as far as possible, there is no generally accepted definition of what seaweed meal is. No standards have been laid down for it. Its quality varies with the type of seaweed used, and with the way in which it is harvested, processed and stored. Most of all, it varies according to the season of harvesting. There is, for example, evidence which suggests that seaweed harvested in May, in Europe at least, is as superior to seaweed harvested in September as May grass is to September grass; and it seems reasonable that this should be so.

A second factor is brought out by Harald Nebb and Arne Jensen in their paper on seaweed-fed dairy cows and bacon pigs. 'There are, of course, many reasons for the conflicting results reported in literature on the value of seaweed meal in rations for domestic animals,' they say. 'No effects, or negative effects, are to be expected when seaweed meal is added to carefully balanced and concentrated diets, since the inclusion of seaweed will reduce the calorific value of the ration. However, when seaweed meal is used as the source of minerals and vitamins, one should foresee beneficial effects, provided that the seaweed meal really contains these components. It is extremely important to secure that the seaweed meal tested satisfies the analytical requirements, and one should always keep in mind that the content of vitamins will decrease upon storage of the meal.'

Finally, one general comment on the use of seaweed for livestock by Dr. Black of the Scottish Seaweed Research Institute. It appeared in the journal *Agriculture* in 1955.

'With pigs,' Dr. Black says, 'seaweed can be introduced as up to ten per cent of the basic diet without affecting the growth rate or the quality of the bacon. With high-yielding dairy cows, ten per cent of their concentrate ration can be replaced by seaweed, without any noticeable effect on their milk yield or butterfat content, although on many farms eight ounces of seaweed meal a day during winter have resulted in a marked increase in milk yield and butterfat. With day-old chicks, up to five per cent seaweed can supply all their vitamin A and D requirements, and with laying hens ten per cent of the basic diet can be replaced by seaweed without any detrimental effect. When seaweed has been fed to laying hens and dairy cows, marked increases in iodine content of eggs and milk have been obtained.

'Since noticeable effects have been obtained by the introduction of as little as one per cent of seaweed in the diet, beneficial results would appear to be due to trace elements, vitamins, or growth-promoting substances active at low concentrations.'

# 11

# THE EVIDENCE OF FARMER AND HORTICULTURIST

I have made a number of claims for seaweed, but they are all based on our experience at Blankney or Holdenby, or on the experiences of others. As far as our own experiences are concerned, I shall have something to say later when dealing with our own garden. The experiences of others are to be found throughout this book.

In order to show that there is sound basis for the claims we make, however, I have interviewed a number of representative users of seaweed products, who have kindly allowed me to pass on what they have told me. Their statements are given in this chapter.

Most of the seaweed products made in this country are used for feeding livestock, and in nursery and horticultural work. I have, however, purposely interviewed one or two outside these main fields in order to give as wide a cross-section of users as possible.

### *Dairy farmer*

One man convinced of the value of seaweed meal when added to animal feedingstuff is Mr. Jeffery Waggett of Cwmconnell, St. Dogmaels, Cardigan. His Friesians once suffered from an outbreak of infertility, whose cause could not be diagnosed in spite of every veterinary test. Mr. Waggett was almost desperate, until he read in the farming press of the value of trace elements in seaweed as a cure for infertility. 'As a final measure I put the affected milking herd on a course of seaweed meal—and the trouble cleared up almost miraculously,' he told me. 'I continue to add a little seaweed meal with the cattle food, and will continue to do so as long as I have the herd.'

### *Forester*

Thus one forester who has used liquid seaweed extract reports 'amazing' results in his tree nursery. He is Mr. T. Rook, who at the time I interviewed him was head forester of the National Trust for the northern parts of Kent, Surrey, Sussex and Hampshire. His headquarters were then at Polesden Lacey near Leatherhead, where he dealt with all aspects of forestry, including planting new areas from the nursery stage, and the thinning and selling of timber. He controlled some 10,000 acres of woodland and common in all.

One important part of his work was raising seedlings in the nursery at Polesden Lacey for the whole of the south-eastern area of the National Trust. This nursery, just under 1 acre of sandy loam on chalk, contains shrubs, seedlings of all kinds of exotic spruce, as well as seedlings of European larch, Norway spruce, Douglas fir, Scotch pine, *Sequoia wellingtonia*, *Thuya plicata*, western red cedar, Lawson cypress, Norway maple, and beech and elm.

The first stage in producing new trees, Mr. Rook told me, was sowing seed in the middle of April. If the seedlings were big enough they were 'lined out' in the following March; if not, this was done a year later.

'At one time,' Mr. Rook told me, 'it was seldom possible to line out in under two years. Now, as a result of giving them liquid seaweed extract, we can line out most varieties at one year. They are much stronger, and produce healthier, cleaner and straighter trees. At one time, after planting out, transplants used to go into a check. Now they just romp away.

'The roots of seedlings treated with seaweed extract are absolutely marvellous. There is much more fibre on them, particularly small hair fibre, and when we lift them at two years for final planting, we usually find they have developed a really good root system. In this condition they go right ahead after being planted out. At one time spruce, in particular, used to go yellow, and stay like that for a year before it began to pick up again. Seaweed treatment stopped all that.

'It also seems to have made the young trees less liable to be attacked by disease. In many nurseries, western red cedar is attacked by a disease called *Keithia thujina*. We used to have it really badly. I wrote to the Forestry Commission when the attacks first took place, and they said the only remedy was to wait for three years and try again. But when I gave the seedlings undiluted seaweed extract, it stopped the disease. We've never had it since, simply because we've used the extract ever since.'

Mr. Rook used to pour undiluted extract on the seed beds from a watering can the day the seed was put in. He gave them a second dose, this time diluted with an equal amount of water, in the middle of August. 'I always make a point of putting this second dose on when I'm practically certain it's going to rain,' says Mr. Rook. 'There's not much point in wasting some of it by putting it on when the weather's fine and hot.' A third and final application might then take place in the second year, when any trees not lined out in the previous year were transplanted.

Mr. Rook told me he ended up with more seedlings fit to plant as a result of using seaweed. It was also usual for larch seedlings, in particular, to reach a height of 15 in. after two years, when previously the average height for these seedlings had been 9 in.

The soil at Polesden Lacey is of a type which responds well to seaweed extract. It is a free-draining soil, but even during the long dry summer of 1959 the tree seedlings suffered no ill-effects.

*Asparagus breeder*

A soil even freer-draining than the sandy loam over chalk at Polesden Lacey is the white fen soil on which Captain A. W. Kidner, the asparagus breeder, first tried out liquid seaweed extract.

Captain Kidner, who has written what I regard as the best book on asparagus growing, was at that time farming one of the most extraordinary soils in England: alluvial chalk on top of peat. This peat, thick with bog oaks, lies above the mineral soil which forms the bed of the southern Fens of eastern England. When the Fens were formed by land subsidence, the forest trees, their roots waterlogged by flooding, fell to the ground and were preserved in peat formed by decaying reeds, rushes and water grasses. In this particular area of the Fens, to the west of Lakenheath, a large lake also formed, which was fed with chalk-bearing water from areas to east and south. In time the bed of this lake—it was called Redmere, a corruption of Reedmere—became thick with precipitated chalk. When the land was drained, this chalk, fine as flour, formed the surface soil of the area. Ploughing has since mixed this chalk with the peat underneath, to form a soil which is roughly 60 per cent chalk, 40 per cent peat.

When Captain Kidner first read our advertisements for hydrolized seaweed extract he did not believe a word of them—as he told us in no uncertain terms at the Royal Show in 1957. But we somehow managed to quell his doubts, for he bought a can of seaweed extract from our stand, and took it home.

'My foreman', he told me afterwards, 'is a good gardener. He took some of the stuff, mixed it in a watering can, and watered certain individual plants in his garden. Within ten days the difference between treated and untreated plants was obvious.

'At that time I was growing about 6 acres of runner beans on white (chalk) fen. I watered one row of these with solution from a watering can. The difference between this row and the others was obvious. I was sold on seaweed.

'And now that I've moved, and am growing asparagus on the light blowy sand of Breckland, just to the east of the fen, it's quite clear that asparagus treated with seaweed extract keeps its foliage green for at least two weeks longer than that in adjoining fields which has not been treated. This not only prolongs the asparagus season, but also prolongs the growing life of the crown itself. Since the crop in spring depends almost entirely on food stored in the roots during the previous summer, the fortnight's extra greening means that the period during which the plant is building up its next year's growth is extended. The plant starts off the next season with that much advantage.'

Captain Kidner began farming on his own at Stoke Holy Cross, Norwich, in 1907. He had been breeding asparagus as a hobby for twenty-five years before he began breeding it commercially on white fen near Lakenheath in 1932. Now he has 52 acres under asparagus, including 20 acres of seedbed, on the sandy Breckland soil which runs down to the fen at Lakenheath. He uses hundreds of gallons of liquid seaweed extract every year.

Seeding takes place at the end of March. The seedlings are sprayed 'as soon as they have developed enough to absorb enough'—which in practice is about the middle of June. Five gallons are given to each acre in 40 gallons of water, and the dose repeated in July or early August.

So much for the seedbeds, where the young asparagus is raised. Cutting beds, where fully developed plants are grown, get 6 gallons of extract to the acre in 40 gallons of water. Spraying on these beds takes place when the plants' foliage is fully developed, usually in July, on stems produced after cutting ceases about 21st June.

Long before Captain Kidner had confirmed our faith in the value of seaweed extract for asparagus on a commercial scale, keen gardeners used to come to our stand at the Chelsea Flower Show to tell us what wonderful asparagus they grew with the help of seaweed extract. We thought then that this was because asparagus, originally a seaside plant, benefited from the salt in seaweed. There can, however, be little doubt that the matter is more complex than that.

*College gardener*

One professional gardener who reports good results with liquid seaweed extract is the head gardener of Jesus College, Cambridge, Mr. W. D. Howard. Visitors to the college often comment on the hanging baskets in the cloisters, with flowering plants and *Nepeta glechoma* trailing down a matter of 9 ft. from the baskets. These basket fuchsias, ivy geraniums and hanging begonias, with three varieties of *Campanula—alba, mayii* and *isophylla*—are all raised on seaweed extract.

Mr. Howard also uses extract on his late chrysanthemums. 'I water them three times a week, from final potting to flowering, with a solution of 2 oz. of extract in 3 gallons of water. This produces quite remarkable foliage, particularly on such late decorative chrysanthemums as 'Yellow Symbol' and 'Flame Symbol'. The judges at the 1964 Cambridge Chrysanthemum Society Show said they had never for years seen two vases of chrysanthemums with such magnificent foliage as those we entered—and I had not, either!'

Bedding plants at the college are also treated with seaweed extract. The peat used in potting soil, and in pricking-off boxes for seedlings, is soaked in a 2 oz. to 4 gallon solution before the compost is made up. 'I have found that recovery on pricking-off is remarkable when seaweed extract is used,' says Mr. Howard. 'No matter what peat I use, I soak it in this solution—and the fibre pots used for bedding out, too.'

Mr. Howard believes seaweed solution gives geraniums outstanding colour and a sound rooting system. 'You can actually see the green colour "rising up" in the geraniums after they've been given seaweed,' he says. 'We water them with solution after they have struck, at the rate of 1 oz. of seaweed extract to 3 gallons of water. We begin potting at the end of November and the beginning of December; and then continue to water them weekly with solution right up to the end of March, this time with a solution of 2 oz. of extract in 3 gallons of water.'

*Nurseryman*

Among many professional nurserymen who use seaweed products, Mr. J. Temple of Bluegate Nurseries, Pyrford, Surrey, is worth quoting for two reasons: first because he sells, not to wholesalers, retailers, or the general public, but to other market growers, whose standards are high; and because he looks on seaweed products not primarily as fertilizers but as pest and fungus controllers.

Mr. Temple's nursery is on flat Thames valley land, a silty loam over silty clay. He describes it as 'very light, but at the same time, glutinous'. The clay subsoil 3 ft. or 4 ft. down will not allow water to drain away, so he is forced to overwinter his outdoor cropping on raised beds. He has some 22,500 sq. ft. of glasshouses and 3,500 sq. ft. of dutch lights, cropped on the three-bed method. This means that each of his twenty-seven ranges of glass has three areas of land to rotate over—backwards, forwards and sideways—in 9 ft. wide bands. In addition he has 14 acres of open land.

In 1938 Mr. Temple had an office job in the City of London. After trying unsuccessfully to join the Navy, he worked for a market gardening firm for three months to gain some experience, and then took over some derelict land not far from his present nurseries, to which he moved in 1940. He now concentrates on growing lettuce, cabbage, celery, marrows and cauliflower plants for other growers, mostly those who specialize in outdoor vegetables grown without glass. He also produces celery, cabbage and lettuce for the wholesale trade. He has a staff of four.

'My business depends on producing good, healthy plants, and as early in the year as possible,' he tells me. 'I must give my plants a boost to bring them on early—and that's why I didn't get very far with liquid seaweed extract at first. It didn't force the crop. Now that I can obtain a grade with added nitrogen, however, it not only gives a nice green colour to anything it's put on, and encourages good sturdy growth—but it also produces the early plants we want.

'But valuable as seaweed extract with nitrogen is, you can't use it on young seedlings except with extreme caution, and in great dilution. I use it at the weakest of strengths on the earliest seedlings—say ¼ gallon of extract for 35,000 cauliflower seedlings. This means that when the solution is made up, there's only the merest suggestion of colour in the liquid. As the plants grow I increase the strength, so that the same number of well-grown cauliflower plants get 1 gallon of extract in solution.

'Seaweed extract with nitrogen is first rate, particularly as far as pests are concerned. We grew 70,000–80,000 celery plants for market in 1964, and in spite of the fact that no pesticides of any kind were used, there was no fly and no blight. We had fly at first, but they just disappeared, as if they didn't like the taste of the seaweed. You could use D.D.T. which builds up, or you can use seaweed, which doesn't. Which is better in human terms?

'Powdered seaweed also seems to be effective against black leg in brassicas, which is a pretty serious complaint. You can have a whole crop destroyed by it. There's a chemical treatment which is quite effective, but seaweed seems to do much the same thing without using any chemical at all.

'But it's got to be put on well before the attack takes place. You've got to say to yourself, "That's where my plants have to go," and then spread the seaweed meal on the soil, and let it work in for two or three months before seeding cabbage or cauliflower. If the results are disappointing in the first year—and they may not be—then you must carry on. Results are bound to be better in the second or third year. And it's not the first, or even the second year, that matters in this direction: it's the progressive build-up.

'It's no good putting down a strip of liquid extract or seaweed meal, and expecting the sort of results you get with ammonia. What you've got to aim at is giving the plants an environment in which they can grow healthily—and clean ground is essential when you're growing plants year after year on the same soil.

'Producing healthy plants is essential to my business. I have to make sure that the man who buys my plants also grows a healthy and successful crop. The conditions in which my plants have to grow may be terrible. If they are, then my plants will have to fend for themselves. They can't do this well if they're "chemical" plants—plants which have relied on chemical spraying for their resistance to pests and diseases. What I want to see is a built-in resistance in the plants themselves, for it's on such things that repeat orders depend.'

Mr. Temple gives all his land under—or going under—glass heavy dressings of dung, and 4 tons of seaweed meal an acre every three years, at least two months before the season's planting takes place. Seaweed meal alone, he says, produces negligible results, at least in his own nursery. But there is a special reason for this. He believes seaweed meal cannot be exploited unless the soil is rich in soil bacteria; and while this may be the case in most cultivated land, his has until recently been an exception. Until he took it over, his nursery was unreclaimed old woodland, with a soil low in organic content.

Mr. Temple believes that if soil has been dunged in the past, then seaweed meal will be effective, even if applied on its own. The organic content is already there. 'But as far as my own land is concerned,' he says, 'seaweed meal won't work without bacteria in the soil. I don't think that dead soil which has been starved and worked out will benefit from seaweed meal unless some sort of bacteria-stimulant like

9. Collecting cast weed, North Berwick, Scotland. By courtesy of Institute of Seaweed Research, Inveresk.

10. The leek on the left was raised on seaweed extract.

dung is added at the same time—all of which means, that to the extent that seaweed meal is food for soil bacteria, to that extent it must fail if the bacteria aren't there to eat it.

'That is why I seemed to be unsuccessful at first with seaweed meal, when I first came to this nursery. It was short of bacterial content, and needed dung to set the cycle of fertility going. Since then we have been getting better crops through using seaweed, and the soil has been getting better too. We have been dressing the ground with seaweed meal at a heavy rate for up to eight years, and the results have been gratifying: consistent, even, and healthy crops, with real vigour and sturdiness.

'Even so, I still look on seaweed products primarily as pest and disease deterrents. The pests and diseases which afflict other people's crops just seem to fade away in the face of seaweed. Fungus troubles such as wire stem in cauliflower and cabbage; yellowing of spinach; rot in leeks, and all the troubles lettuce are heir to—they all seem to fade away in time if you use seaweed.'

Mr. Temple listed for me the pests and diseases which he believes seaweed has helped to control on his nursery, and I give them here:

Cabbage stem weevil
Flea beetle
Club root
Celery fly
Celery leaf spot
Lettuce aphis
Grey mould
Downy mildew
Black leg

With all of these Mr. Temple has had results—and in the case of club root and grey mould, outstanding results. 'Gradually, progressively,' he says, 'seaweed is eliminating these pests.'

One final way in which Mr. Temple's experiences confirm findings elsewhere is in relation to frost resistance. 'Last year and the year before,' he told me, 'our cauliflower plants were grown on land which hadn't been treated with seaweed. We lost both crops from frost. This year, which was just as frosty, we seaweeded the land thoroughly, pricked them off into pots, and now have a fine crop.'

*Apple grower*

A commercial apple grower who also finds seaweed products valuable as pesticides is Mr. J. C. Habgood of Stock, Essex, who has 5 acres of dessert apples, mostly Cox's Orange Pippin grown as dwarf pyramids. The orchard was planted some twelve years ago, and Mr. Habgood relies entirely on biological control to deal with pests and diseases. No chemicals are used, either as artificial feeding compounds or as foliar sprays to control pests and diseases.

A system of feeding is practised whereby stable and farmyard manure are first composted with extra straw, and grass or weed cuttings, before being applied to the soil. This composted material supplies food to the soil bacteria, which in turn break down the material itself. In so doing they release plant foods in a natural form, readily assimilated by the growing trees.

In addition Mr. Habgood gives 10 cwt. of seaweed meal to the acre. He says that it is interesting to see the speed with which this meal is taken down into the soil once it has been reconstituted by rain or moisture. Similar results have been noted by Mr. Habgood when he gave liquid seaweed extract experimentally at 6 gallons to the acre after harvest. He found that the autumn leaf carpet beneath the sprayed trees disappeared into the soil more quickly than that beneath the unsprayed trees, suggesting that the seaweed spray may have made the leaves tastier to earthworms and soil bacteria.

Seaweed extract is also used during the growing season. Beginning at pink bud stage, 4 gallons diluted with 50 gallons of water are applied to the acre. Mr. Habgood believes that this application gives the trees some protection against several degrees of frost. A petal-fall spray of half this strength is followed by a more concentrated application of 5 gallons of extract to the acre in 50 gallons of water. This is given at fruitlet stage.

This third application coincides with the hatching of the winter eggs of red spider mite. Mr. Habgood believes that this spray, and subsequent sprays at the rate of 2 gallons of seaweed extract an acre given at three-week intervals, help to prevent build-up of the mite. Mr. Habgood also believes that these sprays, with his pruning technique, almost entirely prevent attack by the mildew mycelium.

As far as scab is concerned, he points out that the high population of earthworms is capable of dealing rapidly with autumn leaf fall. After the leaves have been taken down into the soil, the overwintering spores of the fungus are destroyed, and the spring inoculent reduced to a low level.

The absence of fungicide sprays has enabled mycelium life to develop, and control lichen and moss satisfactorily. As a result, the trees are virtually free of these growth inhibitors—the small amount remaining provides winter quarters for overwintering predators.

Woolly aphis is present, but is kept under control by the Syrphid larva, as well as the larva of the lacewing, and the specific *Aphelinus mali*—this last being most beneficial in a warm summer. Other forms of aphis, red spider, apple and dock sawflies, codling, tortrix and so on, are held below pest proportions by the large predator population which has built up over the years.

The high standard of health in Mr. Habgood's orchard was confirmed when in 1962 it was judged the winning orchard in Class Two of the County Orchard Competition. It took all the awards, including a trophy for the orchard with the best pest and disease control—given by a chemical manufacturing company.

*Parks manager*

One parks superintendent who has carried out much valuable practical research with seaweed products, particularly in the composting of seaweed meal and the treatment of sports turf with seaweed, is Mr. L. C. Chilcott, parks manager of the new London borough of Brent. Readers may be familiar with the district under its more usual name of Wembley.

Mr. Chilcott has under his charge twenty-eight football pitches; half a dozen rugby pitches; twenty-six cricket tables; seven bowling greens; a number of grass tennis courts; and some 240 acres of sports ground attached to schools in the area—a total of 1,200 acres. The soil is nearly all clay; even the best land has only 8 in. of loam before clay is reached.

Mr. Chilcott's experiences with sports turf are worth describing in some detail. He has always believed that organic substances are essential in cultivation of any kind, and this belief has been strengthened by his observation of cut grass.

'Most turf', he says, 'gets no more organic material than flown grass—grass cuttings which are allowed to lie where they fall. But even if you mow turf once a week, you'll seldom find that flown grass builds up. If you pull the blades of growing grass back a few days after mowing, you won't find even a trace of the dead foliage. The same thing applies even if mowing is less frequent, and the amount of flown grass spread on the turf is greater.

'I can't believe that this absence of organic matter is a good thing,' says Mr. Chilcott. 'It suggests the presence of bacterial activity on a large scale—and also suggests that the bacteria could deal with more organic matter if they had a chance. This means, in a word, that neither they, nor the turf, get all the organic matter they need. I came to this conclusion some years ago, when I decided that our sports and ornamental turf would be improved by the addition of organic fertilizer. I felt like trying seaweed meal because it is slow in decomposing—and therefore long-lasting. I also felt it might be less expensive to mix seaweed meal with other fertilizers as their organic content, rather than to buy complete fertilizer whose organic content had already been provided at greater cost by the manufacturer.

'I was at this time paying over 50s. a cwt. for best-quality complete fertilizer, and this was too high a price to make possible the treatment of all the turf under our control. All I could do with such a complete fertilizer was to treat the actual areas used—bowling greens, cricket tables, tennis courts, football pitches, and so on. The rest had to go untreated. This meant there were areas of sports ground which never did get fertilizer—and they showed it. But if I was to treat the whole of the grass areas of the sports grounds, then I had to find something more economical.

'This I found in a mixture of seaweed and nitro-chalk. (Nitro-chalk is an artificial fertilizer made of ammonium nitrate and chalk, used to give plants nitrogen.) I varied the amount of nitro-chalk in the mixture according to what seemed to be the nitrogen needs of the grass, so that to 3 cwt. of seaweed I would add anything from 28 lb. to 1 cwt. I knew the seaweed meal was slow in decomposing, and that during the period of decomposition the soil bacteria would make demands on the nitrogen in the soil. I also knew that unless this was replaced by a ready source of nitrogen, the grass would suffer from nitrogen starvation. So while I added nitro-chalk to the mixture to help the grass on a short-term basis, my main concern was really to provide nitrogen to go with the seaweed, and replace that being used up by the soil bacteria in their work of decomposition.

'We started using this mixture of seaweed and nitro-chalk in 1958 or 1959, giving at least 3 cwt. of seaweed meal to the acre, plus however much nitro-chalk was thought necessary. It was about twelve months before we saw any change in the grass. But as time went on, and we applied the mixture once and sometimes twice a year, we began to see a change taking place in all our sports areas.

'At one time we used to have tremendous empty patches on our sports fields—areas almost completely denuded of grass—at the end of the winter season. But as time went on these patches became smaller, or the amount of grass which remained in them became greater. And today our worn patches finish up the season in quite reasonable condition—a thing which has been commented on favourably by people from all over the country.

'The fact that well-worn patches come through the winter in better condition results from three things: a strengthening of the grasses already in the turf; a tendency for the types of grass to change; and the fact that some growth, however little, probably continues throughout the winter. The tendency for types of grass to change is seen in two ways. Strong-growing grasses such as perennial ryegrass seem to develop a dwarf type of growth which helps to produce a thicker mat of turf; and one particular variety of grass known as *Poa annua* comes through the winter outstandingly well.

'In the ordinary way, *Poa annua* is thought to be one of the worst grasses for turf. It is coarse and tufty, and for this reason is ruination to bowling greens; but in spite of this, it's not generally regarded as a strong-wearing grass. But under seaweed treatment, *Poa annua* seems to become tougher, wears much better, and comes through the winter well, so that when combined with perennial ryegrass mixtures it helps to produce quite a thick turf. Indeed, we now have football pitches in use from the first days of August which contain large areas of *Poa annua* in combination with other grasses. If, as seems possible, *Poa annua* has become a good servant instead of a bad master as a result of using seaweed, then I won't mind how much of it we have in our sports turf.'

It also seems that treated grass continues to grow throughout the winter, except when the ground is frozen. For this reason Mr. Chilcott feeds the turf even in winter. 'I know that grass can grow in temperatures of 38° F. and above,' he says, 'and I reasoned that if the grass grew during the winter, then I should help it do so. It's mostly nitrogenous fertilizer I apply in winter. I know it's not usually thought sound policy to give turf nitrogenous fertilizer after August, but I think the fact that seaweed is there makes all the difference. Even though the seaweed meal has been applied months before, it won't be completely decomposed; and by absorbing some of the inorganic nitrogen in the nitrogenous fertilizer—which it makes available again in organic form once it has turned to humus—the seaweed reduces the risk of that lush growth in grass which predisposes to disease. In this particular way, and indeed generally, seaweed serves as a safety valve, by helping to control some of the less satisfactory effects of too much inorganic nitrogen in the soil.

'One disease which too much nitrogen encourages is Fusarium wilt; but it is noticeable how minor the attacks of Fusarium in our bowling greens have been since we treated them with seaweed meal.

'Snow mould is another disease which can be encouraged by excess nitrogen, although its main cause is snow lying for a long time on top of the turf. Ground in this country seldom freezes below a depth of 8 in., and at lower depths its temperature is relatively high, and varies very little. In cold weather there is a tendency for this warmth to escape through the surface of the ground. But if there is a long-lying blanket of snow on the surface, then this prevents the warmth escaping, and contains it about the roots of the grass. And it is here, in a snowy season, that snow mould, and indeed other fungus diseases, can be incubated.

'In the bad winter of 1962–3, for instance, our bowling greens were under snow for long periods—as were greens all over the south-east of England. But while dozens of neighbouring clubs could not begin play until weeks after their normal opening time because their grass had suffered from snow mould, we were able to open our greens at the normal time—the last week-end in April. I think most of the credit for this must go to seaweed meal.

'Finally, I do believe that treated turf on our sports grounds is greener than average, and suffers less in drought than was the case before seaweed meal was used. I think this must be because once decomposition has passed a certain point, gradual and continued decomposition of seaweed meal applied over a long period provides the grass with a constant source of nitrogen. It doesn't produce it on too generous a scale, and what it does produce is organic. And there's always some more coming along. Seaweed applied three years ago still has a slight residual effect, that applied two years ago a greater effect, that applied one year ago the greatest effect of all; and since there will always be some seaweed meal more decomposed than the rest, nourishment will always be available for the grass so treated.'

We have already seen that seaweed meal is slow acting because soil bacteria take weeks or months to break it down into humus. We have also seen that during this period of decomposition the bacteria borrow nitrogen from the soil, which is not then available to plants. Thus when Mr. Chilcott, once having proved the value of seaweed meal on grass, tried its effect on pot plants, he found that some plants were tolerant of seaweed while it was decomposing, and some were not. It was this intolerance, caused by the temporary unavailability of nitrogen, which caused him to try composting the meal before use. This he has now managed to do with the help of dried blood, which is one of the richest sources of organic nitrogen. His invention forms a valuable addition to our practical knowledge of seaweed use.

Two lb. of dried blood, when mixed with 1 cwt. of seaweed meal, will quickly cause the mass to rise in temperature and begin decomposing. In summer this begins within forty-eight hours, at other times of the year within a few days. It is best composted under cover to keep out rain, although a sprinkling of water now and again helps decomposition. After four to six weeks of strong activity the composted meal can be used instead of hoof and horn meal with soil, leaf mould or peat, sand, lime, potash, nitrogen and phosphate—and indeed any other accepted constituents—to make soil compost for pot plants. If composting continues for three months the results will be even better. Little weight is lost by composting seaweed in this way. One 4-cwt. load, composted for six months, had lost only 58 lb. in weight when re-weighed—a reduction of from 448 lb. to 390 lb.

Mr. Chilcott uses 2 lb. of composted seaweed meal to every bushel (approx. 40 lb.) of soil and organic material in his compost. If, for example, he makes up a compost of one part loam, three parts peat, one and a half parts sand and one part charcoal, he adds composted seaweed according to the weight of loam and peat in the compost, taking no account, for this purpose, of the sand and charcoal.

Mr. Chilcott did not tell me, although I know from my own observations, that the pot plants he raises are necessarily subject to the most varying and trying conditions—from near-ideal conditions in glasshouses, to the most testing conditions of heat, tobacco smoke, and artificial light at receptions, balls and other functions. Yet even in these trying environments his plants rival, in appearance and condition, those to be seen in the big marquee at the Chelsea Flower Show. His department has also been awarded both gold and silver medals by the Royal Horticultural Society for exhibits at Chelsea and at the Royal Horticultural Society's Hall in Vincent Square, London.

Mr. Chilcott's department uses liquid seaweed extract, although not so extensively as seaweed meal. He told me that it was particularly useful for pot plants, or seedlings in danger of becoming starved before being planted out in spring or summer. The growth it then encouraged, he said, was quite remarkable. Liquid seaweed extract also had another advantage: it could be applied with the chemical sprays used against weeds. In practice he adds it to the 2, 4-D type chemical sprays he uses for selective weed control on lawns, football and cricket pitches, and other grass areas.

## 12

# OUR OWN EXPERIENCES WITH FRUIT AND FLOWERS

Our garden at Holdenby is 400 ft. up, and slopes towards the east. Late frosts are common, and we are subject to east winds in February and March. The underlying rock is ironstone—the dark, mustard-coloured sandstone containing hollow nodules of iron, which has in the past been used for building in Northampton, Kettering, Wellingborough, and other Northamptonshire towns and villages. The soil varies from medium-heavy clay to medium loam, with a pH of between 6·5 and 8. The ground water, or that part of it which finds its way into our well, is one of the hardest our analyst has ever tested.

When we came to live here in April 1952, the garden had received little attention, and certainly no nourishment, since the war. The central lawn, with five long flower beds bordering it, had been kept clean; but the rest of the garden was a wilderness of enormous laurel trees up to 25 ft. high, worn-out apple and pear trees up to seventy years old, and a surrounding barrier of oak, lime and pine trees, so arranged that no view from the downstairs windows was possible. The former vegetable garden was thick with perennial weeds such as nettle and bindweed, and the tennis court was covered with brambles.

Our first task was to fell some of the timber. This gave us more light, and opened up the view from the ground-floor windows—while letting in more of the east wind at the same time. But the £200 we received from the sale of the timber helped to finance the rest of the reclamation work.

It was then a matter of tackling the beds, and the overgrown area of garden, with every weapon which came to hand—weedkillers, brushwood killers and rotary hoes. The weedkillers and brushwood killers were only partly successful; it was persistent and unremitting toil with spade, fork and rotary hoe which did the job. At first we had no help, and could do little more than push forward the margins of our cultivations a few yards each year. After six years, however, we were able to afford a full-time gardener; and now, in 1964, the whole garden is under control save for half an acre of dense laurel, poplar and holly shrubbery.

Our first major change was to turn the old orchard into a vegetable garden is under control save for half an acre of dense laurel, poplar apple and pear trees. We were reluctant to do this, because some of the trees occasionally bore good fruit. On the other hand they were old, and their bark was set and corky. Experts in fruit culture might have been more successful than we in grafting new types on to these old trees; our efforts had little result.

There were other reasons why we felt a drastic reduction in the number of apple and pear trees was necessary. We thought it essential to become familiar with a wide range of garden subjects: if we were in addition to master the highly specialized techniques of fruit growing, it would be at the cost of time spent in vegetable growing. Such a course would also mean duplicating part of the knowledge of fruit specialists already on the staff of our company. To equal, or surpass, their knowledge and experience would be impossible. These were some of the reasons why we decided that we could not, at the same time, be both vegetable growers and orchardists.

There was a further reason why we decided to turn the orchard into a vegetable garden. Had we tried to create a new orchard on the site, or indeed anywhere else in the garden, we would have been fighting against heavy odds. The garden backs on an ancient spinney of some 4 acres, made up largely of oak trees; and experience in our Warwickshire garden taught us how difficult it was to keep clean any orchard overshadowed by mature oak trees. We thought then, and all experience has since confirmed our belief, that the really successful orchard must not only be thoroughly accessible for spraying and pruning, but well away from wild trees too. For these reasons we decided not to try to grow top fruit, but to concentrate on vegetables and flowers.

Once we had turned the orchard into a vegetable garden, we were able to turn part of the original vegetable garden into a rose garden. This we did in 1961, with the enthusiastic co-operation of Mr. Harry Wheatcroft, the rose grower, who recommended varieties to plant, and suggested how the garden might be arranged.

That part of the old vegetable garden which lay beyond the new rose garden we turned into two large island beds, one on each side of a grass path; and here we set out to grow as great a variety of hardy herbaceous plants as we could. Among those in the beds at the time of writing are:

TABLE 35

Plants at Holdenby

*Acanthus*
*Achillea*
*Agapanthus*
*Anthericum liliago*
Alliums in variety
Anemones in variety

Bergamot
Brooms in variety

Campanulas in variety
Chrysanthemums

*Doronicum*
*Dianthus*
Delphiniums
*Dicentra formosa*
*Dicentra spectabilis*
*Deutzia gracilis*

*Echinops*
*Epimedium*
Ericas in variety
Erigerons in variety
*Erinus*
*Eryngium*
Euphorbias

Fuchsias
Foxgloves

Gaillardias

Haberleas
Hebes
Hellebores in variety

Heucheras
*Hydrangea*
Hypericums

*Kerria japonica*
*Kolkwitzia amabilis*

*Lantana*
*Lamium*
*Liatris*
*Lilium candidum*
*Lilium regale*
*Lythrum*
Lewisias
Lupins

*Malva alcea*
Michaelmas daisies in variety

Penstemons in variety
Poppies in variety
Paeonies in variety
Phlox in variety
*Physalis*
*Primula denticulata*

*Romneya coulteri*

Salvias in variety
*Spiraea douglasii*

Verbascums
*Viburnum carlesii*
*Viburnum fragrans*

Yuccas

One particular flower, sweetpea, deserves individual mention because it is the special domain of our gardener, Mr. Maurice Greaves. Before he came to us in 1958 we never had the patience to give sweetpeas the full treatment. But luckily Maurice was already something of a sweetpea expert, and he has now built beautifully symmetrical cages of cane and wire up which to train these flowers.

At one time we tried sowing sweetpea seed in autumn for planting out in the following spring; but mice invariably got inside the frames, and ate the seed. We now soak it for twenty-four hours in a 1 in 400 solution of seaweed extract at the beginning of February; sow singly in fibre pots to save root disturbance; and then water weekly with 1 in 400 solution until planted out.

A further change we made was to build a rock garden. We found space for this by cutting back a 10 ft. wide yew hedge which divided the garden into two more or less equal parts. This hedge we cut back to the uprights, first on the east side, then on the west. We did it with misgiving, and the result nearly justified our fears. It looked awful. We were told by an authority on these things that we had been too drastic, and might lose the whole hedge. But in the event, of the twenty stems which made up the hedge, we lost two only.

It was in the space thus created, on sloping ground east of the hedge, that we put our rock garden. We were at some advantage here, for buried all over the garden are pieces of limestone masonry which formed part of the original palace of Holdenby built by Sir Christopher Hatton, Elizabeth's Chancellor, in 1565, and demolished early in the nineteenth century. These we used as the basis of the rock garden.

It soon became clear that we needed a supply of stone flags to give working access to it, and to provide stepping stones elsewhere in the garden. We were not at first successful in our search. The part of Northamptonshire in which we live produces no flat stone. But then we had a stroke of luck. On one of our holidays in the Pennines, we stayed in a recently modernized farmhouse. All the stone flags from passages, kitchen and scullery were dumped in the garden. We bought 3½ tons of these for £10, and for a further £13 had them put on rail for Holdenby. Some of the flags now give access to the rock garden, and make it possible to work there easily on hands and knees.

The rock garden, where we grow over a hundred kinds of alpines as well as other plants, is now the site of the larger part of our collection of dwarf and slow-growing conifers. Of these we have something more than sixty, either in the rock garden or in our collection of stone troughs. Liquid seaweed extract gives these conifers good root growth, and helps them to withstand long periods of drought, without encouraging them to make more growth than is necessary.

Long before these changes took place we had decided the principles on which we would base our gardening operations. They were these:

(1) Although prepared to use weedkillers, we would use no fungi-

cide or insecticide of any kind. This principle we have never departed from.

(2) We would introduce no manure or fertilizer of any kind, apart from our own seaweed products, and compost made in the garden itself. This principle we have since modified, to make possible comparative trials between seaweed and competitive products. We have also, from time to time, tried mixing seaweed products with other substances, such as dried blood and urea. None of these experiments has been encouraging enough to warrant continuance, at least for gardening use. (We do market seaweed extract with urea, but not for gardening use; and the dried blood used in composting is used with seaweed meal, not liquid seaweed extract, and as an activator, not as a fertilizer.)

(3) We would attempt no plot experiments save on the lawn, where different sections might be subject to different treatments. This principle we have respected throughout.

(4) We would, in time, put up more glass for the controlled observation of seed germination and propagation, and for carrying out pot work generally. This decision, again, has been put into effect.

It was in the rose garden that our work began to produce dramatic results. In the first year we had a good deal of mildew and aphis attack. There were times when some of the bushes looked a little rough. But all we did was spray the bushes with seaweed solution at the first sign of attack, and add seaweed meal to the compost given the beds in autumn.

In the second year we had no serious aphis trouble, and the only varieties attacked by mildew were one or two known to be subject to it, such as 'Chantre'. Thus long before the value of seaweed sprays against aphides and mildews had been proved by experimental work here and overseas, we were able to say that in our conditions at least, the general use of seaweed could prevent insect and fungus attack.

The aphides came, of course; but after sitting around for a day or two they made off. If they delayed, a quick spray with seaweed solution speeded them on their way. For this purpose we found an ideal sprayer in the Dutch knapsack motor-powered machine made by Machinefabriek de Kinkelder, Nijverheidsstraat 2, Zevenaar. This machine can spray a ¼ acre rose garden in twenty minutes.

The original glasshouse, partly below ground, contained a magnificent Madresfield Court vine, so called because the variety was first hybridized at Madresfield Court in Worcestershire. It was subject to mildew, but the grapes had a beautiful flavour, and the skins were not too tough. We carried out trials with this vine for a number of years, giving it varying amounts of seaweed solution. We found the best method was to dig four or five pint-sized holes round the rooting area outside the house; and fill these in mid-spring with a 50-50 solution of extract in water. Later on, when the leaves were half formed, we sprayed the whole house with solution at the same strength.

In this way we increased the yield of fruit considerably. But the house was old, and impossible to make wasp-proof. If we covered the grape bunches with muslin to keep the wasps away, the grapes were attacked with mildew; if we left them uncovered, the wasps ate them before they were properly ripe. For these reasons we cut the vine down. It continues to send up shoots every year, so that we are able to give cuttings to friends interested in viticulture.

Our tomatoes are in a new Crittall glasshouse. They have given us good crops of fruit for eight years, and the crop growing there at the time of writing is as clean as the first we grew, in spite of the fact that the soil has never been changed or sterilized. No commercial grower would think it possible to grow tomatoes in the same soil continuously for eight years without sterilization. We think it is only the use of seaweed which makes this possible. The tomato beds are given 6 oz. of seaweed meal a square yard in the autumn, and we treat the plants with liquid seaweed extract all through their life. Seed is sown in compost soaked in a 1 in 300 solution of seaweed extract, and seedlings and plants watered with a 1 in 300 solution every seven or ten days.

One objective we have set ourselves is to grow as great a variety of shrubs as possible. The fact that we man our own stand at three shows a year, the Chelsea Flower Show included, gives us a chance of seeking the newest specimens on the market. We try to pick shrubs which are pleasing to look at, as well as new to grow; and believe that all the main groups of shrubs are now represented somewhere in our garden.

Our first task with shrubs was to find the best way of giving them a good start in life. At first we used to dig holes, put in a handful of seaweed meal, then a layer of compost, and put the shrub's roots on top. But in rainy weather the seaweed meal became wet and jelly-like, and formed a cold, sticky, water-retaining puddle beneath the plant's roots. As a result we lost several of our purchases. We finally abandoned this technique for the one we describe in the next paragraph, which we recommend for all subjects, whether shrubs, trees or herbaceous plants.

If the shrub or plant or tree is delivered before we have time to plant it, we stand it for a day or two in an old iron wash copper half filled with peat soaked in a 1 in 300 solution of seaweed extract. This improves the roots, and helps them to get away more rapidly when planted. In some cases we do this even when immediate planting is possible. But whether or not the shrub is so treated before planting out, one must then carry out the following routine:

(1) dig a hole deeper than is needed for the plant's roots,

(2) put a good handful of peat soaked in 1 in 300 seaweed extract at the bottom,

(3) plant the shrub in the usual way, and

(4) sprinkle a handful of seaweed meal on top of the soil round the rooting area.

If the soil is dry, a few pints of a 1 in 300 solution of extract may also be watered round the shrub.

Our vegetables, like all other plants in the garden, are given seaweed only, in the form of meal or liquid extract. The results have been excellent, with one exception: cauliflowers. Although we have friends, and customers, who raise excellent cauliflowers on seaweed, we have never produced first-rate cauliflowers at Holdenby. It may be the soil. It may be our own method of cultivation. But with this crop alone we have not succeeded.

All other vegetables, with this one exception, thrive excellently with no manure or fertilizer—only seaweed. Our potatoes are 'Dr. Mackintosh', 'Epicure' and 'Majestic', and our brussels sprouts 'Jade Cross'. We also grow spring, summer and autumn cabbages, broccoli, spinach, broad beans, French beans, runner beans, culinary peas, asparagus, beetroot, carrots, artichokes both globe and Jerusalem, sweet corn, marrows, pumpkins, tomatoes, cucumbers, lettuce,radish, horse-radish, spring onions, parsnips, raspberries, loganberries, strawberries and gooseberries.

The vegetable garden is too big to water, even in the driest weather. The only liquid extract the vegetables get is when certain varieties, such as asparagus, runner beans and peas, are sprayed. We spray asparagus in July to encourage a longer green life, and runner beans and peas in spring to feed the plants through the foliage, and help the flowers to set, particularly in a cold, dry spring. We have proved the value of this technique two or three times at Holdenby.

In autumn the whole of the vegetable garden is given a top dressing of seaweed meal at the rate of 4 oz. to the square yard. Crops such as asparagus, which are valuable enough for such treatment, get from

6–8 oz. Compost is given, irregularly, from our home-made heaps.

The first new lawn we made was laid out in the dry summer and winter of 1961. We prepared the ground with perhaps unusual thoroughness. We dressed the seedbed with 4 oz. of seaweed meal to the square yard, then sprayed the grass with a 1 in 300 solution of liquid seaweed extract as soon as it was about 1 in. high, and had grown enough to benefit.

The success of this technique was proved when we dug up a series of sods five weeks after sowing, and found roots to a depth of over 9 in. We believed that grass with such root development would not suffer from drought, and our confidence was justified. Even though the first real summer rain did not fall until the following June, the lawn never suffered a setback.

We have, of course, a compost heap. Like most domestic heaps, which are seldom turned with the regularity prescribed by most experienced (or, come to that, inexperienced) gardening writers, ours is never turned at all. We add an occasional layer of seaweed meal to the other variegated material it contains, such as vegetable waste, weeds, lawn cuttings, dead rats, and so on. Then, when the heap reaches the top of the wall, we water it with a solution of liquid seaweed extract. This is all the treatment it gets before being dug up in the winter, and spread over the beds in the form of a fertile-looking, friable black soil, swarming with active red worms and soil bacteria.

Some of our most interesting gardening experiences have been with lime-tolerant and lime-intolerant plants. Indeed, the whole question of lime tolerance and intolerance began to fascinate us as soon as we began to take gardening seriously. We had little difficulty in confirming that heathers, heaths, rhododendrons and azaleas do not like lime, although pinks and carnations do. What puzzled us, apart from the reasons for tolerance and intolerance, were borderline plants said by some authorities to be lime-tolerant, by others to be what might be called lime-neutral. Concerning the inclinations of *Lewisia*, *Astilbe* and *Pernettya*, as well as some of the lilies and some of the Asiatic primulas, any extensive gardening library will give completely contradictory information; we therefore set out, not only to study the ways of known lime-intolerant plants, but also those of such twilight specimens as I have mentioned above. In this study we have been helped by the use of the chelated seaweed extracts earlier described. The idea for these came while listening to colleagues discussing chelates at Biarritz in 1961. We then decided to try making up seaweed extract with chelated iron; and after our works chemist had experimented with different strengths, that with an iron content of 1 per cent by weight of standard 'Maxicrop' was adopted. This gave an iron content amounting to rather more than 10 per cent of the dry matter of our ordinary seaweed extract. We gave away scores of ½-gallon cans of this seaweed-with-iron extract to friends, and persuaded some of our orchard customers to try it on chlorotic trees—which they did with immediately satisfactory results.

We have also used it in our garden. In our first experiment we potted up lilies, *Azalea indica* and dwarf rhododendrons, in compost with a pH of 8. In the ordinary way these calcifuges would have died in soil so alkaline; and since the only help we proposed giving them was a monthly feed of 1 in 200 'Maxicrop' with iron, it seemed at first as if the test might be too drastic. When the trial began in October the azaleas and rhododendrons certainly dropped their leaves smartly, and the lilies began to look unhappy. But spring found all the plants growing vigorously; and it was possible to transfer the whole trial in May to our stand at the Chelsea Show, with specimens in the alkaline compost as vigorous, lively and floriferous as those in compost with a pH of 5. In fact the *Lilium auratum* given 'Maxicrop' with iron had brighter colours than plants of the same variety growing in peaty compost.

In a second series of experiments we planted lilies, azaleas, dwarf rhododendrons, lewisias, lime-hating gentians and *Lithospermum diffusum* 'Heavenly Blue' in ordinary garden soil with a pH varying from 7·5 to 8, and watered their roots monthly with a 1 in 200 solution of extract and chelated iron. All the treated calcifuges responded, but none so dramatically as the lithospermum. However this plant may behave elsewhere—and many people say they grow it perfectly well in limy soil—it looks miserable, attenuated and chlorotic if left to itself in our garden at Holdenby. But after one dose of iron-chelated extract, the lithospermum greened up beautifully within a matter of three or four days.

It is clear, if writers on the subject may be trusted, that intolerant varieties respond differently in different soils, even when the pH of the soils is the same. For this reason my observations, which are reliable as far as Holdenby is concerned, may not be reliable with soils of the same pH elsewhere. But four years' experimenting make us think that gardeners with similarly alkaline soils who want to grow calcium-intolerant plants should not be disappointed if they go about it as we have done.

One point I should make here—although it results only partly from

11. Sheep grazing seaweed on the shore of North Ronaldshay, Orkney, Scotland (John Topham).

12a. Seaweed on drying walls, Orkney, Scotland. By courtesy of Alginate Industries Ltd.

12b. Unloading air-dried seaweed, South Uist, Scotland. By courtesy of Alginate Industries Ltd.

our experience at Holdenby—is this: shrubs and other plants which need chelated-iron extract to grow in alkaline soil can be given it in solution watered round their roots. With orchard trees, on the other hand, watering the soil in this way is useless. Apples, pears, plums and other fruit trees suffering from chlorosis must have their iron-chelated seaweed extract given in the form of a foliar spray. It should be given in the period May to June in four or five doses, at a total rate of 12–14 gallons an acre. I have seen no case of lime-induced chlorosis in fruit trees which has not shown some response, and sometimes a dramatic response, to this treatment.

The second chelate we had made up was with iron, magnesium and manganese. It is intended for use with plants suffering from magnesium deficiency. So far we have used this in our own garden on camellias and azaleas only, but the results with camellias have been dramatic. These have always been a difficult subject with us; but now, for the first time, we are growing sturdy plants with leathery, glossy green leaves, plenty of buds, and flowers which do not drop off.

The fact that we grow so many different subjects in our garden inevitably means that we know something about a lot of things, although those about which we know a great deal are few. But calcifuges are one subject about which we have gained first-hand practical knowledge. We now know that soil with a high pH, in thoroughly healthy organic condition, will support calcifuges better than soil of exactly the same pH with a lower organic content, and less bacterial life. We know, too, that soil which is becoming more acid grows calcifuges better than soil of exactly the same pH which is becoming more alkaline. In this respect, at least, we think we have added something of value to the knowledge of how garden plants behave.

APPENDIX A

# RECOMMENDATIONS FOR INDIVIDUAL PLANTS AND CROPS

This section contains recommendations for the use of liquid seaweed extract and seaweed meal. It will be understood that as far as liquid extract is concerned they refer only to the product of my company, 'Maxicrop'. I am sorry to introduce this commercial note once more, but for our liquid extract, which is based on complete alkaline hydrolysis of the plant under pressure and heat, and is not simply an aqueous infusion of macerated seaweed, I must claim specific qualities. The nature of its manufacturing process means that 'Maxicrop' retains constituents and qualities denied extracts not made by alkaline hydrolysis. While 'Neptune's Bounty' is simply the registered name of the seaweed meal we sell, and we claim for it no more, and no less, than can be claimed for any good commercial seaweed meal on the market, 'Maxicrop' is a patented manufacturing process as well as the product which results from it. The claims we make for it must be specific, and our recommendations can apply to it alone.

These remarks apply generally to all the observations in this book unless otherwise made clear. They must apply with even more force now that it is a question, as here, of making recommendations. For the sake of ease in reading I shall not use our trade names consistently throughout this section. Where I am dealing with liquid seaweed extract, however, my recommendations must be understood as applying to 'Maxicrop' alone. Such recommendations as are made for the use of seaweed meal apply to any good commercial meal on the market.

Liquid extract can be used with advantage on any farm or horticultural crop, although in practice economic and other factors usually restrict its use to grass and cereals, top fruit, soft fruit, potatoes and other root crops, peas and beans, hops, celery, brassicas and kale. In horticulture it is chiefly used on tomatoes, cucumbers, carnations, chrysanthemums, roses and flowers in general, as well as lettuce and all pot plants. Dilutions recommended vary from 1 in 100 for lettuce to 1 in 1,000 for certain pot plants. They are detailed under the name of the crop. Whatever the dilutions given, it is nearly always possible, although for other reasons not always desirable, to give stronger doses without harming the plant. It is not always advisable, because weaker solutions are usually more effective and, as far as cost of material is concerned, cheaper. The cost of labour will of course have to be considered when increasing the number of times spraying takes place; but in practice, better results are always obtained by giving the same amount of extract twice, in greater dilution, rather than once, at greater strength.

'Maxicrop' is compatible with all pesticides if used fresh. If it is used in continuous watering systems, a dilution of 1 in 1,000 is recommended.

'Maxicrop' with chelated iron may be given weekly to the soil round calcifugeous plants in a 1 in 400 solution. If given as a foliar spray, a 1 in 200 solution may be given once a fortnight. Orchards and soft fruit suffering from severe iron deficiency may be given a solution containing equal parts of 'Maxicrop'-with-iron, and water. This should be given once a fortnight, at the rate of 2 or 3 gallons to the acre, until a total of between 12 and 14 gallons have been given. 'Maxicrop'-with-iron is not compatible with all pesticides.

'Maxicrop'-with-nitrogen contains 17 per cent organic nitrogen in the form of urea. It is recommended for use (for example) on pasture grass to give a quick nitrogen boost, and as a post-harvest spray for apple trees. It is compatible, freshly made, with all pesticides. It is, however, intended for use by commercial growers only, and should be used by amateur gardeners with extreme care—if at all. (We use the word 'gardener' here and later for those who do not earn their living by growing, 'grower' for those who do.)

As far as standard liquid extract is concerned, general rules for the amateur gardener are: water the soil surface with a 1 in 300 solution of extract when sowing seeds, planting out annuals, or transplanting new shrubs. If the soil is too wet, soak roots of bedding-out stuff in a 1 in 300 solution after lifting from seedbed or seedbox, and before putting in final position. Soak all easily handled seeds in a 1 in 400 solution, and cuttings in a 1 in 300 solution, before sowing or planting, for twenty-four hours. If water supplies allow, water beds in dry weather with 1 in 300 solution, and spray growing plants with 1 in 100 solution, whenever you feel inclined. Add extract to all compost heaps in a 1 in 200 solution, whatever the material being composted.

As earlier remarked, seaweed extract may be given in greater dilutions in the United States.

As a general rule, seaweed meal should be used on gardens in autumn or early spring, at the rate of 2 oz. a square yard for lawns, and 4 oz. a square yard for all beds except those containing heavy clay and light sand, which should be given from 4–6 oz. Subject to what has already been said in the chapter on the application of seaweed products, seaweed meal should not be dug in, but given as a top dressing. It should not be put on beds in which seedlings are being pricked out. This is important. As we have already seen, soil bacteria will borrow nitrogen from the soil while breaking down the seaweed, and thus starve the seedlings of a nutrient essential for their growth. This does not apply to legumes, which have their own stores of food, or to mature plants which already have their own root ball to nourish them.

Seaweed meal at the rate of 4 oz. a square yard may be put at the bottom of trenches dug for potatoes, sweet peas, and other crops. For intensive cropping, heavier dressings of up to 8 oz. a square yard may be given. (These dressings may be reduced in the United States.)

Seaweed meal is, of course, applied to the ground only. Since it is slow in acting, and long lasting, it is supplied to the soil at long intervals, usually once, and never more than twice, a year. Unless composted, it is applied in its original state. In this it differs from liquid extract, which is always given in a solution of water, and sometimes in a solution of water and pesticide. Extract is applied to the plant itself even more than to the soil, and at intervals which may be as frequent as once a week.

For these reasons it is necessary to give more detailed instructions for the use of liquid extract than for the use of seaweed meal. These are set out below, first for general, and then for individual, varieties. Where there is no qualification, standard 'Maxicrop' is intended. Where 'Maxicrop' plus iron or nitrogen is indicated, this is made clear.

Cereals should be given one dose of from 4–6 gallons an acre of extract between the period late March to June, while the cereal is growing. If 'Maxicrop' with nitrogen is used, then eight parts of water, which may or may not include other spray material, should be used with one or more parts of extract. If it is necessary for management reasons to give fewer sprays, then dilutions as strong as one part of extract to one of water may be given—but only if standard extract, not that with added nitrogen, is used. Strong dilutions of standard extract may cause temporary scorch, but are unlikely to result in that reduction of leaf area which follows permanent scorch.

Cereals which suffer a setback in late June as a result of a cold or dry spell, or both, will be given renewed vigour by a single application of extract at the strength given in the previous paragraph. A single application in June will also give the finishing touches to a malting barley—I say malting barley because it would be uneconomic to give it to other, less valuable, barleys. Such applications will also produce stiffer straws and increased yield—although, as we have already seen, the cost of spraying cereals with extract may make the operation uneconomic.

On some soils, nitrogen top dressing can be halved if 2–3 gallons of extract plus nitrogen are given—say once for cereals and potatoes, more often for blackcurrants.

Pasture grass should be given extract as a foliar spray from late March onwards. Four to six gallons to the acre can be given in dilutions of one part extract to eight of water. The water may contain other spray material, and the solution should be given twice, at most, between March and June, according to the strength of solution. If fewer applications make life easier for the farmer, he can give standard extract neat—or extract with nitrogen in solutions no stronger than one in eight—without risk of permanent harm. Since standard extract is compatible with all ordinary commercial sprays of which we have knowledge, mixing it with weedkiller will prevent the setback which chemical weedkillers sometimes cause in grass.

Standard extract should be used on grass with suspected mineral deficiency, and on swards rich in clover. When application takes place early or late in the season—for early bite or second crop—extract plus nitrogen may be used. On alkaline soil, extract plus iron should be used.

## FRUIT

Top fruit and soft fruit should be given 6–8 gallons to the acre for the whole season according to the size of the trees. Three or four applications of equal parts of extract and water will be necessary according to season and climate. The first spray should be given when there is enough leaf area to absorb the solution, usually in the third week in May.

Soft fruit can be given three sprays of 50-50 solution, the first when the leaves are out, a second when the fruit trusses set, and a third after harvest. This final spray delays leaf fall by two or three weeks, and strengthens next season's growth. If the soil is watered, the solution should be weakened to 1 in 200. An autumn dressing of seaweed meal may also be given round the roots of soft fruit trees.

These are general recommendations for fruit. Specific recommendations follow.

*Apples*

Spray with standard extract at blossom time and fruit set, in a 50-50 solution with water, or water and pesticide. Give 8 gallons to the acre for the total season, that is, 1 gallon an acre in each spray given. If extract with nitrogen is used, a minimum of 8 gallons to the acre should be given in about 60 gallons of water, at any time between early October and a fortnight before leaf fall. This treatment gives extra nitrogen to help the tree produce fruit buds in the following spring.

Fruit spurs should contain at least 2 per cent nitrogen to produce good fruit. Extract plus nitrogen provides this at a time when all other tree processes are competing for it. The period after early October is the best time to apply it to the leaves. It is then that leaf nitrogen moves back into the stems in preparation for leaf fall.

Apple trees which show chlorosis should be given 8–10 gallons of extract with iron an acre in four or five doses between May and June, in 50-50 dilutions.

*Blackcurrants*

Give bushes up to four years old a total of 6 gallons to the acre in two separate sprays: the first when spraying for big bud mite, the second at fruit stage. Give up to 6 gallons of extract an acre for waist-high trees, up to 8 gallons for shoulder-high trees, both in 50-50 solutions. Blackcurrants of any age may be given extract plus nitrogen to help bud formation in the following spring, on the same terms as are outlined above under *Apples*.

*Cherries*

As *Apples*.

*Peaches*

As *Apples*. To cure leaf curl in all but the most obstinate cases, spray fortnightly with 1 in 10 or stronger solution.

*Pears*

Spray with standard extract in a 50-50 solution with water, or water and spray solution, at blossom time and fruit set. Total amount for two sprays, 6–8 gallons of extract to the acre. Pears showing chlorosis should be given a total of 8–10 gallons an acre of extract with iron, in four or five doses between May and June, in 50-50 dilutions.

*Plums*

As *Apples*.

*Raspberries*

Treat as for blackcurrants. A post-harvest spray of extract with nitrogen may be given to strengthen new growth on terms outlined under *Apples*.

*Strawberries*

Spray twice with standard extract, first when new leaves are developed, then when the flowers show colour. Three gallons should be given to the acre at each spraying, making a total of six for the season. Dilutions here are unimportant. The extract can be applied neat, or diluted to suit the sprayer used. A post-harvest spray of 3 gallons to the acre can also be given to strengthen the crowns for the following season.

## VEGETABLES

The following are recommendations for individual vegetables.

*Asparagus*

Dress beds in autumn with 4 oz. of seaweed meal to the square yard. Water beds in early spring with a 1 in 300 solution, and spray foliage in July with 1 in 300 solution, or stronger, to build up roots for next year's crop. (These are recommendations for private gardeners.)

*Beans, broad*

Spray crop with 3 gallons an acre in a 50-50 solution when plants are 4 in. high, and with a further 3 gallons an acre at the same strength two weeks later.

## *Recommendations for individual plants and crops*

*Beans, dwarf and runner*

Give 2 gallons an acre in a 50-50 solution as climbing tendrils appear, followed by another 2 gallons an acre at the same dilution two weeks later.

*Brassicas*

Give a total of 6 gallons to the acre in two or three applications as soon as there is enough leaf surface to absorb the spray. The sprays can be in 50-50 dilutions, and may be given every two to four weeks.

*Brussels sprouts*

Plants in seed beds should be sprayed once with 4 gallons an acre in a 50-50 solution, immediately after transplanting.

*Cabbage*

See *Brassicas.*

*Cauliflower*

See *Brassicas.*

*Celery*

Self-blanching celery should be given 6 gallons an acre in a 50-50 solution on two occasions in the period July–August, at two- to four-week intervals. Wide rows of celery should be given 2 gallons an acre, applied to the rows, and at the same dilution, twice in the period July–August.

*Cucumbers*

The soil in which cucumbers are rooted should be watered every seven to ten days after the seedling stage with 1 in 400 solution. If the cucumbers are rooted in straw bales, extract plus nitrogen may be given to the bales in a 1 in 8 solution of water, or other spray preparation, to help break down the bales. The nitrogen in the spray can then also help to satisfy the plants' nitrogen needs.

*Hops*

A total of 6 gallons an acre should be given as a foliar spray with copper wash, or other spray, in three separate applications of 2 gallons each. These can be given in dilutions of 1 in 10 between May and August. Further applications can be given at any time to help bines overcome a weather setback.

*Kale*

Give a total of 6 gallons to the acre in two or three applications, as soon as there is enough leaf surface to absorb the spray. Dilute to 1 in 10, and complete spraying cycle by June.

*Leeks*

These vegetables can be made resistant to seed-borne virus if the seed is soaked in a 1 in 400 solution of extract for twenty-four hours, and then dried on blotting paper before being sown. The growing crop can be watered as for brassicas.

*Lettuce, cold frame or cold house*

Water the soil two or three times with a 1 in 100 solution during the effective growth of the lettuce.

*Lettuce, field*

Spray soil, or plants, or both, with a 1 in 500 solution before hearting up.

*Peas*

Water or spray garden peas with 1 in 200 solution every ten days when tops show above ground, soaking seed for twenty-four hours in 1 in 300 solution for quicker germination. Market garden peas should be given 4 gallons an acre in two sprays.

*Potatoes*

Gardeners should soak uncut potato sets in a 1 in 200 solution of extract for twenty-four hours before planting and scatter 4 oz. of seaweed meal to every square yard of trench bottom. Farmers should spray the rows with 3 gallons an acre of solution just before final earthing up. Then give a second 3 gallons an acre as late as possible before the leaves close over the rows. Dilution can be anything between 1 in 4 and 1 in 20 in each case.

*Root crops generally*

As *Potatoes.*

*Spinach*

Spray once with a 1 in 10 solution, 6 gallons to the acre, as late as possible while access can be had to the crop.

*Tomatoes*

If seedlings are pricked out into peat cubes, these should first be soaked in 1 in 100 solution. No feeding will then be necessary before planting out. The soil should then be watered with 1 in 400 solution every seven to ten days or, if watering is continuous, with 1 in 1,000 solution. A foliar spray with 1 in 100 solution should also be given in the middle of the season to make sure the plants do not suffer from mineral deficiency while cropping heavily. If tomatoes are rooted in straw bales, extract plus nitrogen may be watered over the bales in a 1 in 8 solution of water, or other spray liquid, to provide food for the bacteria which break down the straw. The nitrogen in the spray also helps to satisfy the plant's nitrogen needs (which gardener and grower will assess, in the case of this and other crops, by the appearance of the plant).

Tomatoes grown by ring culture (in which a bottomless 10-in. pot containing soil and plant rests on a bed of aggregate, gravel, sand, ash, or other stable water-holding material) can be given a special 'tomato' extract containing nitrogen, phosphorus and potassium. This provides the extra nutrient needed by a plant which may produce 7 lb. or 8 lb. of food from its small rooting mass of soil. Such extra nutrient is not provided by standard seaweed extract, whose value depends more on the release, than the supply, of nutrients.

Standard extract mixed with proprietary tomato foods may also be used for ring culture.

## FLOWERS

Bedding plants will be stronger if the peat or potting compost is soaked in 1 in 400 solution before use. All flowers will profit if the soil in which they grow is watered every seven or ten days with 1 in 400 solution.

Pot plants should be watered weekly with 1 in 400 solution, although cyclamen, primula and cinerarias respond well to frequent watering with 1 in 1,000 solution.

Seaweed meal should not be used in pot mixtures unless composted, or made up one month before, as described earlier. Used in these conditions, a small amount of seaweed meal in the potting compost helps to retain moisture, to feed the plant, and to buffer it against the over-watering, over-feeding, over-heating, over-lighting—and their opposites—to which most pot plants grown in private houses are subjected.

## *Recommendations for individual plants and crops*

Most people like feeding pot plants. To put the matter at its lowest, feeding them with dilute seaweed solution will do them less harm than anything else. Pot plants which look like dying have even been restored, miraculously on occasion, by a liquid seaweed spray. These instructions seem cynical. I can only say that one of the commonest remarks we hear from those looking at the pot plants we, and others, show at Chelsea is: 'I had one of those. But it died.'

Here are more detailed recommendations.

*Carnations*

Water the stock plant every seven days for four weeks with 1 in 100 solution before cuttings are taken. Then soak cuttings and rooting medium in 1 in 400 solution for better root formation. Once planted, water the soil every seven to ten days with 1 in 400 solution.

*Chrysanthemums*

Water the stock plant every seven days for four weeks with 1 in 100 solution before cuttings are taken; then soak cuttings and rooting medium in 1 in 400 solution for better root formation. Once planted, water the soil every seven to ten days with 1 in 400 solution. 'Maxicrop' plus nitrogen may be given chrysanthemums as a foliar spray in a solution of water, or other spray liquid, not stronger than 1 in 300, to help satisfy the plant's nitrogen needs, or as a soil feed at the same dilution.

*Cineraria*

Water frequently with 1 in 1,000 solution.

*Cyclamen*

Water frequently with 1 in 1,000 solution.

*Daffodils*

Soak compost at planting time with 1 in 200 solution, then water occasionally with 1 in 1,000 when they are brought into the heat.

*Freesias*

Soak seed or compost in 1 in 400 solution for twenty-four hours before sowing in 2½ in. paper pots. Put in 6–10 in. pots when roots come through the paper, and water weekly with 1 in 1,000 solution. Soak compost with 1 in 200 solution before planting bulbs.

*Hyacinths*

As *Daffodils.*

*Primula*

Water frequently with 1 in 1,000 solution.

*Roses out of doors*

Spray two or three times during the season with 1 in 25 solution, beginning when the leaves are about half grown. Increase number of sprayings in a bad mildew or aphis year.

*Roses under glass*

Water with 1 in 400 solution every seven to ten days. Increase strength to 1 in 300 at the end of the flowering season to bring bushes back into condition.

*Sweet peas*

Soak seed in 1 in 400 solution for twenty-four hours before sowing, to improve germination and root formation. This is particularly effective with such difficult germinators as 'Swan Lake', whose seed most growers chip to help germination. If the seed is soaked, chipping will be unnecessary, and germination even quicker. Put 4 oz. of seaweed meal to the square yard in the bottom of the trench. Spray with 1 in 200 solution every ten days.

*Tulips*

As *Daffodils.*

## OVERSEAS CROPS

Many of the crops which profit from spraying with extract are common to temperate, sub-tropical and tropical areas. The recommendations already given for the treatment of such crops in this country apply also overseas, except that applications in the United States can be reduced and that changes in spraying times (for example in the southern hemisphere) may have to be made according to changes of season. For the sake of overseas growers, however, we repeat the instructions for world-wide crops below, varying them where necessary for those in sub-tropical and tropical areas.

In all cases the dose suggested is the total for the whole season. It may be given all at once, but better results will be obtained if it is given in two, three or four instalments.

In order to save shipping costs, we export our extract in powder form for reconstituting with water at the point of use. It dissolves easily. Once reconstituted, it can be applied undiluted on mature plants, and will not damage leaf, blossom or fruit. The use of undiluted extract also makes spraying from the air easier because the load to be carried is lighter. Where time and labour are available, however, diluted extract may produce better results.

In some cases dilution rates will be decided by the type of machine used, since different rates suit different sprayers. As a general rule, however, these are reasonable dilutions for the crops listed: grass, 1 in 8; fruit, 50-50; roots, anything from 1 in 4 to 1 in 20; peas and beans, 50-50. In all these cases, and in those later quoted, the ratios apply to reconstituted extract mixed at the rate of 1 lb. of powder to 1¼ gallons of water.

It is best to spray when the weather is dry and warm—except in the case of grass, where either dry or wet warm weather is suitable.

*Grass*

Four gallons to the acre from first spring growth.

*Fruit*

Medium-sized trees and soft fruit bushes should be given 6 gallons an acre, large fruit trees 8 gallons an acre, starting at pink-bud stage.

*Roots*

Roots, including potatoes, should be given 8 gallons an acre—in the case of potatoes, just before final earthing up, in the case of other roots, as late as possible before the leaves close in over the rows.

*Peas and beans*

Four gallons an acre, as late as possible after the rows close over.

Other crops, not in general grown commercially in Britain, should be treated as outlined below. All may be given extract in a 50-50 solution, or any greater dilution according to time, labour and spraying machine. Maize and sugar may be given dilutions of 1 in 8. Spraying should be done in warm, dry weather.

*Apricots*

Eight gallons an acre when blossom appears.

*Citrus fruits*

As *Apricots.*

*Cloves*

Six gallons an acre when new leaves begin to appear.

*Coffee*

As *Cloves.*

*Grapes*

Before planting out, soak the roots of the young vine-set in 1 in 100 solution. Give 4 gallons an acre as a foliar spray as soon as there is enough leaf growth to absorb it, and before the blossom has set. Then give the same amount when applying a copper sulphate wash.

*Ground nuts*

As *Cloves.*

*Maize*

Four gallons an acre before the leaves have hardened off.

*Passion fruit*

As *Apricots.*

*Peaches*

As *Apricots.*

*Pineapples*

As *Cloves.*

*Sugar*

Four gallons an acre before the leaves have hardened off.

*Tea*

As *Cloves.*

*Tobacco*

Give 4 gallons an acre early in the season, when there are enough leaves to absorb the spray; and a second spray of 2 gallons to the acre when the leaves are half grown.

# *Recommendations for individual plants and crops*

### *Flowers*

The following recommendations for flowers have been made by the Department of Agriculture, Government of Bahrein. They may be of value to others who grow flowers in equally hot and arid climes. All are foliar sprays unless otherwise stated.

The recommendations were made following tests which took place with 'Maxicrop' in 1961 and 1962, and so to some extent the table is not only a summary of recommendations, but also of the results of trials on which the recommendations are based. In some cases, which are noted, extract was given to correct chlorosis. Where no reference to a specific purpose is made, it can be assumed that extract was given as a booster, or simply to produce, or maintain, healthy and colourful plants. Distilled water was used in making up the spray solutions, and all spraying was done in the cool of the evening. Where chelated iron was added, this was Sequestrene 330 F.E. at the rate of from ½–1 oz. for 4 gallons of solution—that is to say, it was not our own already iron-chelated extract which was used, but our standard extract to which a commercial iron chelate was added.

It is not suggested that fertilizer should necessarily be added on the same scale, but the Department of Agriculture records the fact that the trial plots were given 6 lb. of farmyard manure a square yard, as well as 1 oz. of a combined nitrogen-phosphorus-potassium fertilizer.

Improvements, however slight, were reported in all cases save one: with violas the comment was 'Fair results, flowers disappointing'. In the case of *Dianthus*, *Dimorphotheca*, *Linaria*, *Linum*, *Salvia* (blue), stocks and verbenas, it will be seen that the results were excellent.

TABLE 36

Flowers at Bahrein

These recommendations for the foliar spraying of flowers with seaweed extract were made following trials by the Department of Agriculture, Bahrein. Notes on the results of the trials are included.

| | *Time of application* | *Teaspoons of extract per gallon* | *Trial results* |
|---|---|---|---|
| *Ageratum* | 3–4 weeks after germination | 3 (chlorotic correction) | colour and growth improved, good heads |
| *Alyssum* | 3–4 weeks after germination | 3 | colour and growth improved, early flowering |
| *Antirrhinum* | 3–4 weeks after germination | 3 (chlorotic correction) | colour and growth slightly improved |
| | 7 weeks after germination | 3 (booster and correction) | colour and growth improved |
| | 10-day intervals | 3 | colour good, growth steady (sprays given with chelated iron $\frac{1}{2}$–1 oz. in 4 gallons of solution) |
| *Asparagus sprengeri* | 3–4 weeks after germination | 3 | colour and growth improved |
| | after potting | 3 (applied to soil) | no set-back |
| | after 7-day intervals | 2 | steady growth maintained |
| *Aster* | 3–4 weeks after germination | 3 (chlorotic correction) | colour and growth improved |
| | after 10-day intervals | 3 (chlorotic correction) | colour and growth improved, good heads (sprays given with chelated iron) |
| *Calendula* | 3–4 weeks after germination | 3 | increased growth, early flowering, good heads |
| Candytuft | 3–4 weeks after germination | 3 | increased growth, early flowering |
| Carnation | before planting | 5 (applied to soil) | good germination |
| | 3–4 weeks after germination | 3 (booster and correction) | improved growth and colour |
| | 10-day intervals | 3 | growth and colour maintained, good heads (chelated iron added to spray) |
| *Celosia comosa* | 3–4 weeks after germination | 3 (slightly chlorotic correction) | improved growth and colour |
| | 6–7 weeks after germination | 3 | growth and colour maintained |
| *Centaurea* | 3–4 weeks after germination | 3 (slightly chlorotic correction) | improvement in growth and colour |
| | 6–7 weeks after germination | 3 | growth and colour maintained, good heads |
| *Cleome* | before planting | 5 (applied to soil) | fairly good germination |
| | 3–4 weeks after germination | 3 (chlorotic correction) | slight improvement in growth and colour |

| | *Time of application* | *Teaspoons of extract per gallon* | *Trial results* |
|---|---|---|---|
| o | 7-day intervals | 3 (slightly chlorotic correction) | steady improvement in growth and colour (chelated iron added to spray) |
| Chrysanthemum (annual) | 3–4 weeks after germination | 3 | improved growth and colour |
| | 6 weeks after germination | 3 | improved growth and good heads |
| *Clerodendron fallax* | before planting | 5 (applied to soil) | fair germination |
| | 10-day intervals | 3 (chlorotic correction) | improved growth and colour (chelated iron added) |
| *Coleus* | before planting | 5 (applied to soil) | fair germination |
| | 10-day intervals | 3 | good growth, colours excellent |
| *Dahlia* | before planting | 5 (applied to soil) | improved germination |
| | 3–4 weeks after planting | 3 (booster and correction) | improved growth and colour |
| | 10-day intervals | 3 (booster and correction) | steady growth, good colour, large heads |
| *Dianthus* | as for carnations | | excellent results |
| *Dimorphotheca* | as for carnations | | excellent results |
| *Gaillardia* | 3–4 weeks after germination | 3 | improved growth, large flowers |
| Geranium cuttings | before bedding | 5 (applied to soil) | very little die-back |
| | 10-day intervals | 3 | steady growth (chelated iron added) |
| | if potted, 7-day intervals | 2 | steady growth and good flower heads |
| Hollyhock | 3–4 weeks after germination | 3 | improved growth and colour |
| | after transplanting | 3 | no set-back, good growth, large heads |
| Larkspur | before planting | 5 | good germination |
| | 10-day intervals | 3 | improved growth, large heads |
| *Linaria* | as for larkspur | | excellent results |
| *Linum* | as for larkspur | | excellent results |
| *Petunia* | before planting | 5 (applied to soil) | fairly good germination |
| | 10-day intervals | 3 | improved growth, large flowers (chelated iron added) |
| *Salvia* (red) | as for petunias | | fairly good results |
| *Salvia* (blue) | 3–4 weeks after germination | 3 | excellent growth and good flower heads |
| Stocks | as for petunias | | excellent results |
| *Verbena* | as for petunias | | excellent results |
| *Viola* | as for petunias | | fair results, flowers disappointing |
| *Zinnia* | 2–3 weeks after germination | 3 | strong growth and excellent flower heads |

## APPENDIX B

# RECOMMENDED RATIONS FOR STOCK

Seaweed, unless fed directly to animals in seaside areas, is given in the form of meal. This may be eaten as it is, as part of a farm-mixed ration, or as part of a balanced feedingstuff produced by a manufacturer. Both the meal and the balanced rations which include it are available in powder form, or compressed in nuts.

The golden rule of livestock feeding, that all changes should be made gradually, applies to the introduction of seaweed into an animal's diet. Calves, when introduced to seaweed meal at six to seven weeks old, should be given 1 oz. a day only, mixed with their rations. Cattle, if not brought up on seaweed, should at first have only 2·5 per cent of their ration in the form of seaweed meal. This may be increased to between 5·0 and 7·5 per cent over a matter of three to four weeks. If the meal is given to combat infertility, the proportion may be raised to 10 per cent. Many months' feeding on this scale may be necessary before some of the more obstinate conditions are cured.

Pigs should be given only 2·5 per cent of their rations in the form of meal at first. The proportion may then be increased gradually until it reaches 5·0 per cent.

Poultry can take from 5·0 to 7·5 per cent of their rations in the form of seaweed meal, and do not seem to need gradual introduction to it. Baby chicks can begin eating it at once. Balanced poultry rations containing seaweed meal are available in most districts of the United Kingdom, in meal and pellet form.

Recommended amounts for other animals are, for horses, up to ½ lb. a day; for goats, up to 4 oz. a day; and for sheep, up to 4 oz. a day also, depending on the size of breed. In all these cases, seaweed meal should be added to the rations in small amounts at first, and increased gradually until the desired proportion is reached.

## *Recommended rations for stock*

Here are some typical animal rations containing seaweed meal.

*Cattle*

Three typical dairy rations for cows.

TABLE 37

Rations for dairy cows

| | *1* | | *2* | | *3* | |
|---|---|---|---|---|---|---|
| | *cwt.* | % | *cwt.* | % | *cwt.* | % |
| Seaweed meal | 1¾ | 8¾ | 1¾ | 8¾ | 1¾ | 8¾ |
| Oats—crushed or coarsely ground | 3½ | 17½ | 5¼ | 26¼ | 7½ | 37½ |
| Barley—ground | 4 | 20 | 5 | 25 | 3 | 15 |
| Beans (or peas)—coarsely ground | — | — | 7 | 35 | — | — |
| Wheatfeed | 3 | 15 | — | — | — | — |
| Flaked maize | 2½ | 12½ | — | — | 3 | 15 |
| Palm kernel (or coconut cake meal) | 2 | 10 | — | — | — | — |
| Undecorticated cotton (or groundnut) | — | — | — | — | 2 | 10 |
| Groundnut (or soya meal) | 3 | 15 | — | — | 2½ | 12½ |
| Fish meal | — | — | 1 | 5 | — | — |
| Sterilized steamed bone flour | ¼ | 1¼ | — | — | ¼ | 1¼ |
| | 1 ton | 100% | 1 ton | 100% | 1 ton | 100% |
| Starch equivalent | 63·2% | | 63·8% | | 62·0% | |
| Protein equivalent | 14·0% | | 13·8% | | 12·8% | |
| 4 lb. ration supplies: | | | | | | |
| Starch equivalent | 2·53 lb. | | 2·53 lb. | | 2·48 lb. | |
| Protein equivalent | ·56 lb. | | ·55 lb. | | ·51 lb. | |

*Pigs, finishing rations*

Here are three typical pig finishing rations. It is calculated that their total digestible nutrients, and protein both crude and digestible, are as set out at the bottom of the table.

TABLE 38

Finishing rations for pigs

| | *1* | | *2* | | *3* | |
|---|---|---|---|---|---|---|
| | *cwt.* | % | *cwt.* | % | *cwt.* | % |
| Seaweed meal | 1 | 5 | 1 | 5 | 1 | 5 |
| Barley meal | 12 | 60 | 10 | 50 | 9 | 45 |
| Wheat (coarsely ground) | — | — | — | — | 2¼ | 11¼ |
| Oats (finely ground) | — | — | 2 | 10 | — | — |
| Wheatfeed | 6 | 30 | 4 | 20 | 6 | 30 |
| Flaked maize | — | — | 1¼ | 6¼ | — | — |
| Groundnut meal | — | — | 1 | 5 | 1½ | 7½ |
| Fish meal | 1 | 5 | ½ | 2½ | — | — |
| Limestone flour | — | — | ¼ | 1¼ | ¼ | 1¼ |
| | 1 ton | 100% | 1 ton | 100% | 1 ton | 100% |
| Crude protein | 14·6% | | 14·0% | | 14·5% | |
| Digestible protein | 11·2% | | 11·7% | | 12·1% | |
| Total digestible nutrients | 67·0% | | 68·2% | | 67·4% | |

*Pigs, sows and weaners*

In sow and weaner meals the proportions of the constituents are varied, as seen in the table below. If these are fed in winter—or to intensively housed stock at any time of the year—six million international units of vitamin A, and one and a half million of vitamin $D_3$, should be added.

TABLE 39

Sow and weaner rations

| | *1* | | *2* | | *3* | |
|---|---|---|---|---|---|---|
| | *cwt.* | % | *cwt.* | % | *cwt.* | % |
| Seaweed meal | 1½ | 7½ | 1½ | 7½ | 1½ | 7½ |
| Barley meal | 7½ | 37½ | 6½ | 32½ | 8¼ | 41¼ |
| Oats (finely ground) | — | — | 2 | 10 | — | — |
| Wheatfeed | 8 | 40 | 6 | 30 | 8 | 40 |
| Flaked maize | 1¼ | 6¼ | 1½ | 7½ | — | — |
| Groundnut meal | — | — | 1 | 5 | 1 | 5 |
| Fish meal | 1¾ | 8¾ | 1½ | 7½ | 1 | 5 |
| Limestone flour | — | — | — | — | ¼ | 1¼ |
| | 1 ton | 100% | 1 ton | 100% | 1 ton | 100% |
| Crude protein | 16·4% | | 17·3% | | 16·3% | |
| Digestible protein | 13·4% | | 14·5% | | 13·6% | |
| Total digestible nutrients | 66·0% | | 67·8% | | 66·0% | |

*Poultry*

These three mashes for layers are suitable for extensive, or intensive, systems, including deep litter and battery. If used in winter, or for birds intensively housed at any time of year, at least seven and a half million international units of vitamin A, and one and a half million of vitamin $D_3$, should be added to each ton of mash. Flint grit, as well as limestone or oyster shell grit, should be available at all times.

TABLE 40

Poultry rations

| | *1* | | *2* | | *3* | |
|---|---|---|---|---|---|---|
| | *cwt.* | % | *cwt.* | % | *cwt.* | % |
| Seaweed meal | 1 | 5 | 1 | 5 | 1 | 5 |
| Barley meal | 4 | 20 | — | — | 2¾ | 13¾ |
| Wheat (coarsely ground) | 5 | 25 | 6 | 30 | — | — |
| Oats (finely ground) | 3¼ | 16¼ | 5¼ | 26¼ | 5 | 25 |
| Maize gluten feed | — | — | 2 | 10 | — | — |
| Bean or pea meal or grass meal (16% protein) | 1 | 5 | — | — | — | — |
| Wheatfeed | 3 | 15 | 4 | 20 | 6 | 30 |
| Maize meal | — | — | — | — | 2 | 10 |
| Groundnut meal | 1½ | 7½ | — | — | — | — |
| Soya bean meal | — | — | — | — | 2 | 10 |
| Fish meal | 1 | 5 | 1½ | 7½ | 1 | 5 |
| Steamed bone flour | ¼ | 1¼ | ¼ | 1¼ | ¼ | 1¼ |
| | 1 ton | 100% | 1 ton | 100% | 1 ton | 100% |
| Crude protein | 17·0% | | 17·0% | | 16·5% | |
| Digestible protein | 14·5% | | 14·2% | | 14·1% | |
| Productive energy | 80·0 therms per 100 lb. | | 77·4 therms per 100 lb. | | 85·3 therms per 100 lb. | |

APPENDIX C

# SOME SELLERS OF SEAWEED PRODUCTS

Companies in this country marketing seaweed products for use in agriculture and horticulture at the time of writing include:

*Barkers & Lee Smith Ltd.* and *Seaweed Agricultural Ltd., both of Barkers Mills, Lincoln.* (Seaweed Agricultural Ltd., whose foundation is described in the first chapter, is now controlled by Barkers & Lee Smith Ltd.)
*Centurion Agricultural Company Ltd., Oak Park, Dawlish, Devon.*
*Wilfrid Smith (Horticultural) Ltd. of Gemini House, High Street, Edgware, Middlesex.*

'Marinure' powdered seaweed manure, made from *Ascophyllum nodosum.*

'Marinure' N.P.K. seaweed fertilizer, seaweed meal with added nitrogen, phosphate and potash.

'Marinure' Concentrate seaweed spray, made from seaweed fortified with nitrogen, phosphate and potash.

*Grebon Holdings Ltd., Burcot Grange, Abingdon, Berkshire.*

'Alginure', primarily a soil conditioner, made from seaweed meal treated with alkali, water and heat.

*Chase Organics (Great Britain) Ltd., Gibraltar House, Shepperton, Middlesex.*

SM3, sold as 'Sea Magic' in some countries, an all-organic concentrate of seaweeds with about 10 per cent of land plant tissues added. Made by aqueous extraction.

Seaweed is also contained in the products of Pan Britannica Industries Ltd., Britannica House, Waltham Cross, Herts ('Baby Bio', 'Bio Plant Food', 'Bio Humus', 'Bio Plant Builder', 'Bio Lawn Tonic' and 'Bio Compost Maker'); Amoco (U.K.) Ltd., Vigzol House, Eastney Street, London, S.E.10 ('Naturfeed').

APPENDIX D

# RECENT RESEARCH RESULTS

*by Ernest Booth*

---

IN the four years since the first edition of this book was published, research has explained some of the valuable properties of seaweed, and a number of valuable field trials have confirmed earlier results and extended the use of liquefied seaweed into new fields. The fact that these results come from Canada, Chile, Czechoslovakia, France, Russia and the U.S.A., indicates the world-wide interest in seaweed.

Whilst seaweed contains relatively small amounts of nitrogen and phosphorus, experience shows that the use of seaweed brings about a disproportionate increase in soil nitrogen and 'available' phosphorus. The effect on soil nitrogen has been studied in Russia where it was shown that seaweed caused vigorous growth of ammonia-producing bacteria; the soil nitrogen decreased slightly initially, but this was followed by a sharp increase of 200 parts per million. Research at the University of Chile confirmed that an increase in the 'available' phosphate was brought about by seaweed and increased yield and phosphorous uptake by the plant was demonstrated. This research also showed that seaweed chelated with iron in the soil to the extent of up to 40 parts per million, and that seaweed was as good as synthetic iron chelates for the treatment of chlorosis of citrus trees.

Alginic acid was the first soil conditioner discovered by Quastel (1947) and many years later (1963), when the relative efficiency of a number of soil conditioners was tested at Woburn, it was noticed that alginic acid increased both the percentage of seeds which germinated and the yield. (A similar effect had already been noted with seaweed extracts, see pp. 106-9, 132). At this time, alginic acid had been known since 1881, but it had only been found in the brown seaweeds and it was assumed that the acid was peculiar to the seaweeds. Its effect on the soil and seed germination were well established facts, which were to remain a curiosity for several years. Then, in 1964, alginic acid was isolated from a bacterial culture (of a *Pseudomonad*) and two years later it was isolated from a culture of a very common soil bacteria *Azobacter vinelandii*. The bacteria, like all living organisms, absorb nutrients and excrete their waste products. It had been known for many years that the nitrogen excreted by *Azobacter* made a large contribution to soil fertility. The discovery that they also excrete the soil conditioner alginic acid, which also affects seed germination and yield, points to yet another link between soil bacteria and soil fertility. This discovery also puts the manurial value of seaweed into perspective, and removes some of the mystery associated with muck.

In an account given by Blunden at the 1968 International Seaweed Symposium, the influence of some of the constituents of seaweed on the growth of mustard was recorded. The seeds were grown in vermiculite and watered with a complete nutrient solution (a formula recommended by Hewitt) with and without the addition of laminaran, mannitol and soluble alginates. Whilst mannitol produced a slightly significant increase in the fresh weight of the plants, laminaran had the opposite effect. Sodium alginate produced a significant reduction in the weight of the plants at all the concentrations used in the experiment (equivalent to 0·05, 0·1 and 0·2 per cent alginic acid). Surprisingly, potassium alginate, at the same concentrations, caused a slight increase in the fresh weight of the plants at the lowest concentration and had no effect at the other two application rates. It is, in fact, most probable that alginates would complicate this test by precipitating the metal ions in the nutrient solution.

The effect of 'Maxicrop' on the increased shelf life of peaches (pp. 122-3) was confirmed by further trials carried out at Clemson College between 1963 and 1967, when four different varieties were tested. It was shown that spraying early in the season had the greatest effect and all the varieties under test showed extended shelf life. At Prague University, Povolny confirmed that the Norwegian seaweed extract 'Algifert' had the same effect as 'Maxicrop' in decreasing the storage losses of peaches. This trial extended over three years (1969–71) and used four varieties of peach which were sprayed three times at 5–6 day intervals in late July and early August. In each variety, the flesh of the sprayed fruit was firmer, and the losses in an 8-day storage trial was decreased from 55 per cent in the controls to 21 per cent in the fruit sprayed with the seaweed extract. Povolny extended this trial to three varieties of apricot in 1969 and 1970, when the treatment with the seaweed extract reduced the 12-day storage loss from 30 to 12 per cent. A feature of these trials was the comparison of the proprietary seaweed extract 'Algifert' with a simple laboratory extract of seaweed meal which proved quite ineffective; this demonstrates the commercial expertise required to manufacture an effective seaweed extract.

Some years earlier, (1966, 1968) Povolny showed that spraying gherkins with the seaweed extract 'Algifert' not only increased the yield by 42 per cent, but he also demonstrated that the stored fruit were much more resistant to softening and rotting. This result was followed by a cucumber trial in 1971, which showed a yield increase of 17 per cent and an increased profit of 146 per cent. This trial also showed that the treatment with the seaweed extract gave a substantial increase in the fruiting period of the plants and there was a marked decline in the incidence of red spider mite which is invariably associated with the glasshouse cultivation of this plant. Povolny has also published (1968, 1969) two papers on the effect of the seaweed extract on apples when he showed there was no significant difference in the acidity or sugar content of the sprayed fruit (Cox's Orange, Nonnetit and Winter Pearmain), but the flesh was firmer and kept better in store. After storage, '. . . there was a slight increase in the acidity of Cox's Orange which was detectable on tasting. The apples of the control trees were considerably over-ripe, mealy and of low acidity whereas . . . apples from the trees which had been sprayed were crisp and tartishly acid.'

In three separate trials with 'Maxicrop' on strawberries (Cambridge Favourite), the average yield was increased by 20 per cent, and the incidence of botrytis was reduced from 22 to 4.6, 15.4 to 4.8 and 12.9 to 1.7 per cent respectively. As might be expected, storage losses followed a similar pattern and, when samples of the fruit were canned, the treatment was shown by the Fruit and Vegetable Preservation Research Association to have no adverse effect on flavour and taint.

At the 1971 International Seaweed Symposium held in Japan, Blunden reported a number of commercial trials with the seaweed extract 'S.M.3.' A trial with peppers on three one-acre plots gave a 6·3 per cent increase in yield from spraying at 0·5 gal/acre, and

26·6 per cent increase at 1 gal/acre; in each case, the peppers from the treated plants had much better keeping qualities. Another trial, which involved several thousand orange trees sprayed at 1 gal/acre over a period of several years showed a small increase in yield. In 1970, there was an accidental delay in transit, which revealed that the sprayed fruit kept well, whereas the untreated fruit was rotten. Thus pure chance added yet another instance of the beneficial effect of seaweed on the shelf-life of fruit.

Blunden's paper also records many other interesting successful trials covering a wide range of plants. In a trial with bananas, carried out by the Jamaica Banana Board, six-months-old plants were sprayed at four different application rates (0·5, 0·75, 1·0 and 1·25 gal/acre) and, in each case, the banana bunches were formed eight weeks earlier than with the controls. The best results, a weight increase of 12·2 per cent, was achieved with an application rate of 0·75 gal/acre. Leaf analysis showed no detectable difference in the level of the major nutrients (N, P, K, Ca and Mg) but high manganese levels were recorded, and the lower application rates had the greatest effect. This is in line with similar observations made by Franki on several occasions between 1958 and 1964. In a trial in Florida with tomatoes, using 'S.M.3' at 1 gal/acre, a yield increase of 20 per cent was recorded. Potatoes responded similarly; when three acres were sprayed with one gallon of the extract diluted to 100 gallons, the yield of tubers was increased by 37 per cent. A trial with Golden Bantam hybrid sweet corn was carried out in Rhode Island, using the seaweed extract at 0·5 gal/acre. Within ten days, the treated plants were about 25 per cent taller than the controls, the leaves were broader and much darker in colour, and the stalks were thicker. A count of tassled ears on two rows gave a total of 147 for the controls, and 209 for the plants treated with seaweed; the cobs ripened quicker on the treated plants, which ultimately showed an increased yield of 56 per cent.

Similar beneficial results were obtained with maize in a trial carried out in France with 'Maxicrop' in 1971. The plants were sprayed at 1 gal/acre on 10th May and again on 30th May at the reduced rate of 1 gal/3 acres. Harvesting was completed on 15th October, when an increased yield of 26 per cent was obtained, which was mainly due to an increase in the average weight of the cobs.

Some very clear results have been obtained with celery at the West Sussex School of Agriculture using the self-blanching varieties Avonpearl and Lathom. The seed was sown on 5th March, pricked out on 13th April and planted out in the third week in May, spaced at 10 inches in 10-inch rows, in wide span polythene tunnels where the soil had been given a base fertilizer treatment of 120 units N, 80 units $P^2O^5$ and 120 units $K^2O$. Two plots received a further dressing of Nitro-chalk (80 units N/acre) on 15th June, and 'Maxicrop' was then sprayed on one of these plots, and on an untreated plot. The crop was harvested on 2nd August, trimmed for market and weighed with the following results:

TABLE 41

| *Treatment* | *Average weight/head (ozs)* | |
|---|---|---|
| | *Avonpearl* | *Lathom* |
| Nitrochalk | 21 | 22½ |
| 'Maxicrop' | 22 | 25 |
| 'Maxicrop' plus Nitro-chalk | 25 | 26½ |

It is evident that 'Maxicrop' had more influence on the average weight of the heads than the treatment with Nitro-chalk, and the plots treated with Nitro-chalk were improved even further by treatment with 'Maxicrop'. This observation confirms many similar results, which suggest that 'Maxicrop' assists the plant to utilise the vast quantities of fertilizers now used in horticulture.

This effect was reinforced by a trial with blackcurrants carried out in 1972 at Bradenham Hall Farm, Norfolk, by the kind permission of the farm manager, Mr. M. Rowe. The plants (Baldwin) were established in 1964 and the trial plots received a dressing of 112 units N per acre in the spring of 1972. 'Maxicrop plus Nitrogen' and three proprietary foliar feeds were applied on the trial plots at the manufacturers' recommended application rates on four occasions (4th, 18th April, 2nd May, 12th June) and the crop was harvested on 3rd August, when the crop weights were recorded with the following results:

TABLE 42

| *Foliar feed* | *Yield (tons/acre)* | *Cost (£/acre)* | *Increased yield (%)* |
|---|---|---|---|
| Control | 6·86 | — | — |
| 'Maxicrop plus N.' | 8·20 | 6·00 | 19·7 |
| Brand A | 7·25 | 6·75 | 5·7 |
| Brand B | 7·35 | 5·00 | 7·2 |
| Brand C | 7·38 | 4·20 | 7·7 |

Now in this trial, similar amounts of nitrogen were given by each of the four foliar sprays, but the increased yield from the 'Maxicrop' treated plants was markedly better than from the other three foliar sprays. Again, the effect of 'Maxicrop' on the utilisation of the fertilizer by the plant is clearly demonstrated. The only indication of a reason for this effect comes from Russian work, particularly that of Sorokina, who has shown that various trace elements considerably increased the uptake of NPK from the soil, and also increase the translocation of those nutrients within the plant. So far, this work has been confined to melons, cucumbers, tomatoes, alfalfa, oats, wheat and potatoes and, usually, the trace elements boron, copper, zinc and molybdenum have been applied singly. In all the recorded work, these elements have increased the uptake of nutrients by the plant and, in the few instances where mixtures of these elements were used, the effect was greater than that given by any metal used alone. There are indications, therefore, that this increase in the uptake of fertilizer which has often been noticed in trials, is caused by the trace elements in 'Maxicrop'.

Trials do not always give the expected result and may be made invalid by unforeseen circumstances, or an unexpected result may occur but, despite the number of trials involved, not a single unfavourable result has been recorded. Typically, a trial was carried out in 1971 by the Institute of Potato Research, Czechoslovakia, which, disappointingly, gave only a minor increase in yield but analysis showed the starch content of the tubers was increased from 15·8 to 16·3 per cent; well, starch is more nutritious than water and, in fact, a high starch content is demanded by crisp manufacturers. Some disappointment was caused in 1971 in a trial with Brussel sprouts which was organised at Gatehouse Farm, Sussex, with the permission of Mr. C. G. Craggs, who was growing under contract for Birds Eye Ltd. This proved to be a hot, dry summer and Birds Eye suggested that foliar sprays should be applied to all crops, and 'Maxicrop' was included in the programme. In the end, the crop was ploughed in because of a surplus of sprouts that year, and all that could be achieved was a taint and flavour test. The Fruit and Vegetable Research Association reported no taint but enhanced flavour, so not all was lost, even though circumstances nullified the more important aspects of the trial. In its way, a most disappointing trial was planned on a young apple orchard at Watlington Fruit Farm, Norfolk, with the kind permission of Mr. Rockcliffe This started in 1967 as a five-acre trial on a seventy-acre site which had been planted in 1961 with Cox, Worcester Pearmain and Bramley's Seedling. Since the orchard was so recently established, results were not expected for some years but, by 1969, Mr. Rockcliffe and his fruit foreman were convinced they could see sufficient evidence to warrant spraying the whole orchard with 'Maxicrop', and only three rows of trees were left as a control. Thus the trial, as such, was completely invalidated, even though the overall result can only be considered most favourable.

Strawberries are a crop which responds to 'Maxicrop' particularly well and several trials have already been recorded, which show its beneficial influence on botrytis (pp. 119-20). Over the past few years more highly successful trials have been carried out with Cambridge Favourite 442, and this work has been extended to a number of other varieties by trials in Italy and France. One of the trials in England extended over three seasons, and the following excellent results show the great benefit derived from 'Maxicrop' in the unfavourable conditions which prevailed in 1970 and, perhaps even more important, the treatment led to more even cropping from year to year:

TABLE 43

| *Year* | *'Maxicrop' (lb/acre)* | *Control (lb/acre)* | *Increase (%)* |
|---|---|---|---|
| 1969 | 14,193 | 11,920 | 19 |
| 1970 | 13,058 | 5,598 | 133 |
| 1971 | 11,430 | 9,328 | 22 |

Three growers in Norfolk supplied strawberry fields in 1971 for trials with 'Maxicrop'. Each area was given the normal base fertilizer and herbicide treatment, but botrytis control was divided between the fungicides Benlate and Elvaron which, in effect, gave two controls. The plants were sprayed with 'Maxicrop' in mid-April and mid-May, and the fungicides Benlate and Elvaron were applied in early June and were considered as separate controls; 'Maxicrop' increased the yield by 10·0 per cent in one case, and by 14·8 per cent in the other.

A trial with eight varieties of strawberries was started in France, when 200 plants of each variety were planted through black polythene strip in August 1968. In the following years, half the plants were treated with 'Maxicrop', diluted 1:50, at the rate of 45 gallons per acre, on 24th and 30th April, 7th and 21st May; the total quantity of standard 'Maxicrop' was rather more than 3·5 gallons per acre. After harvesting, the treated area showed an average increase of 9·3 per cent and, through earlier ripening and better quality fruit, the income per acre was increased by 14·3 per cent. The response of the individual varieties is shown below:

TABLE 44

| *Variety* | *Yield (lb/acre)* | | *Income (£/acre)* | |
|---|---|---|---|---|
| | *'Maxicrop'* | *Untreated* | *'Maxicrop'* | *Untreated* |
| Merton Princess | 16,544 | 13,552 | 1,250 | 1,035 |
| Senga Preconsana | 9,328 | 9,856 | 645 | 725 |
| Kennedy | 19,624 | 17,864 | 1,523 | 1,295 |
| Gorella I | 10,296 | 9,504 | 849 | 646 |
| Gorella II | 17,688 | 15,576 | 1,478 | 1,215 |
| Albert Ier | 18,392 | 15,752 | 1,317 | 1,132 |
| Senga Gigana | 12,936 | 12,584 | 913 | 843 |
| Vola | 20,504 | 19,976 | 1,422 | 1,327 |
| Average | 15,664 | 14,333 | 1,174 | 1,027 |

This was followed by a trial in Italy which was organised by the Osservatorio per le Malattie delle Plante di Verona, in which 4,800 virus-free Gorella plants from cold storage were planted through black plastic strip on 15th July 1970, into ground which had been dressed, per acre, with 30 tons FYM, 6 cwt. of a 10.10.10 fertilizer, and 10 cwt. of potassium sulphate. The plants were covered with polythene tunnels in January 1971, and Benomyl–50 (0.06 per cent; 150 gal/acre) in seven sprays, starting from blossom initiation, was used to control botrytis; 'Maxicrop' (dilution, 1:100; 1.33 gal/acre) was applied to half the plants at the same time as the fungicide. Whilst there was practically no difference in the total harvest from the two sets of plants, the plot treated with 'Maxicrop' ripened earlier and commanded a better price; the details were as follows:

TABLE 45

| *Picking date* | *Yield (lbs)* | | *Income (£)* | |
|---|---|---|---|---|
| | *'Maxicrop'* | *Control* | *'Maxicrop'* | *Control* |
| 24 April–6 May | 699 | 444 | 142·80 | 84·50 |
| 8–18 May | 1,163 | 1,419 | 169·75 | 208·85 |
| 20 May–3 June | 581 | 551 | 42·45 | 40·50 |
| Total crop | 2,443 | 2,414 | 355·00 | 333·85 |

When these figures are converted to a 'per acre' basis, this trial showed a profit of £180 after allowing for the cost of the 'Maxicrop'.

This series of trials, carried out under a very wide range of climatic and growing conditions, provides abundant evidence for the value of 'Maxicrop' in strawberry growing.

The use of 'Maxicrop' by vine growers is extending rapidly in America, on the Continent and in New Zealand irrespective of whether the grapes are used for dessert, wine making or dried fruit; the usual application rate is four applications of two gallons per acre, beginning as soon as there is sufficient leaf to absorb a foliar spray. The effect on the new growth is very striking; the wood is thicker, well ripened and in great demand for cuttings by the local nurserymen who, incidentally, soak the cuttings in 'Maxicrop', diluted 1:100, to promote rooting. The bunches are more open, which both assists botrytis control and enables the grapes to grow more uniformly. Increased yields are commonly reported, but more notable results are improved colour and increased sugar content, two factors which influence the quality of the resultant wine.

A trial carried out at Champagne Mercier at Epernay, France, by the C.I.V.C. (Interprofessional Committee for Champagne Wine) compared various commercial foliar feeds (F.F.1–5, alone or in combination) with 'Maxicrop' on the variety Pinot Meunier. The grapes were harvested on 27th September 1971, with the following results:

TABLE 46

| *Row No.* | *Treatment 1970* | *Treatment 1971* | *Yield (lb/acre)* |
|---|---|---|---|
| 10 | None | None | 5,580 |
| 8 | None | None | 5,970 |
| 3 | 'Maxicrop' | 'Maxicrop' | 6,470 |
| 4 | 'Maxicrop' | None | 6,350 |
| 2 | 'Maxicrop' | None | 6,140 |
| 12 | F.F. 5 | None | 6,190 |
| 5 | None | F.F. 1 | 6,050 |
| 6 | F.F. 2 | None | 5,340 |
| 11 | F.F. 5 | F.F. 4 | 5,200 |
| 9 | F.F. 2 | F.F. 3 | 5,200 |
| 7 | F.F. 2 | F.F. 2 | 4,750 |
| 13 | F.F. 5 | None | 4,670 |

The two control rows (8,10) show a normal yield variation ($\pm$ 195 lb/acre) and the rows (2,3,4) treated with 'Maxicrop' all showed a significant increase over the controls, whereas only two of the other seven treatments showed a positive result, and five gave much lower yields, some of which can only be described as catastrophic. The C.I.V.C. note that yields in 1971 were only about 60 per cent of the 1970 results, and blame this on the cold, wet spring, when the chemical sprays seemed to reduce flower set. This dependence on weather is a universal hazard for all growers, and the fact that 'Maxicrop' appears to benefit flower set in adverse spring weather is of widespread interest.

This effect of 'Maxicrop' is by no means a new discovery. As long ago as 1967, a Chateauneuf du Pape vineyard won two gold medals at the Paris Concours General Agricole where, in the following year, these wines swept the board, taking the gold, silver and bronze medals, as well as winning two gold medals and two silver at the Concours Vinicole d'Orange. What suits Chateauneuf is good for anything!

# BIBLIOGRAPHY

Books and periodicals to which reference has been made are listed below. Further references may be obtained from the publications of the Institute of Seaweed Research, Inveresk, Midlothian, Scotland, and those of Norsk institutt for tang- og tareforskning (The Norwegian Institute of Seaweed Research), Trondheim, Norway.

*Books*


Landsborough, the Rev. D., *A popular history of British seaweeds*, Reeve and Benham, 1851.

Chapman, V. J., *Seaweeds and their uses*, Methuen, London, 1950.

Dickinson, Carola I., *British seaweeds*, Eyre & Spottiswoode, London, 1963.

Newton, Lily, *Seaweed utilisation*, Sampson Low, London, 1951.

Fogg, G. E., *The growth of plants*, Penguin Books, Harmondsworth, Middlesex, 1963.

Mackean, D. G., *Introduction to biology*, John Murray, London, 1962.

International Seaweed Symposium, 1st, 2nd, 3rd, 4th and 5th, Pergamon Press, Oxford.


*Periodicals and leaflets*


*Agricultural Science, Journal of* (London)
1947 (37) 257

*Agriculture* (London)
1955 (62) 12, 57

*Agronomy Journal*
1964 (56) 444

# *Bibliography*

Alginate Industries Ltd. (22 Henrietta Street, London, W.C.2)
Various leaflets
American Society for Horticultural Science (Michigan), *Proceedings*
1944 (44) 49
*Animal Science Journal*
1959 (18) 836
*Botanica Marina* (Hamburg)
1961 (3) 17, 22
*Canadian Journal of Agricultural Science*
1954 (34) 181
*Canadian Journal of Botany* (Ottawa)
January 1952 (30) 78–97
*Canadian Journal of Microbiology* (Ottawa)
1963 (9) 169
Challenor Society (National Institute of Oceanography, Wormley, Godalming, Surrey)
1959, Mowat
*Chemistry and Industry* (London)
1959 998, 1,376
1962 725
Clemson College of Agriculture (South Carolina, United States)
*Bulletin on liquid and dry seaweed*
Research series 23 and 24
*Commercial Grower* (London)
1st January 1965 29
*Dairy Research, Journal of* (London)
1954 (21) 299
*Economic Entomology, Journal of* (College Park, Maryland, United States)
1964 (57) 95
(56) 503
Edinburgh and East of Scotland College of Agriculture, annual reports
1958 and 1959 references to copper and sheep
*Empire Journal of Experimental Agriculture* (London)
1957 (25) No. 97 51
*Experimental Horticulture* (London)
1963 (9) 39
F.A.O. annual fishery statistics
Fisheries of Scotland reports
1960 79

1961 88
1962 95
*Fishing News International* (London)
1963 2(3) 269
*Garden News* (Peterborough)
12th October 1962 1
16th October 1964 18
*Grower* (London)
7th May 1960 1,056
6th August 1960 256
26th November 1960 1,045
8th September 1962 366
8th June 1963 1,150
15th June 1963 1,198
14th December 1963 1,031
12th September 1964 442
29th May 1965 1,160
4th December 1965 902
*Helminthology, Journal of*
1950 (24) 91
*Horticultural News* (New Brunswick)
May 1964 (Driggers)
*Horticultural Science*
1963 (38) 40
Institute of Seaweed Research annual reports, Inveresk, Midlothian
1961 to 1965 inclusive
Irish Institute for Industrial Research (Ballymun Road, Glasnevin, Dublin)
'Re-growth of *Ascophyllum nodosum*'
Israel Research Council Bulletin
1963 11 D (4) 230
Linnean Society Proceedings
1956 (166) 87
*Marine Biological Association Journal* (London)
1960 (39) 433
'Maxicrop' leaflets and technical bulletins (Maxicrop Ltd., Holdenby, Northampton)
*Nature* (London)
1951 (168) 728
1953 (171) 356, 439
1958 (181) 1,499

1961 (190) 109; (191) 684
1963 (198) 1,282; (199) 389; (200) 453
'Neptune's Bounty' leaflets (Seaweed Agricultural Ltd., Barkers & Lee Smith Ltd., Barkers Mills, Lincoln)
*New Zealand Journal of Science and Technology* (Wellington)
1942 23 Section B 149
*Norway Exports* (Oslo)
1962 62–6
Norwegian Seaweed Institute (Trondheim)
Leaflets dealing with seaweed meal for cows, pigs, sheep, chicks, laying hens and pigs
*Plant Physiology* (Washington, United States)
1963 (38) 175
*Plant Physiology Annual Review* (Stanford University, Palo Alto, United States)
1959 (10) 257–76
*Polar Record* (Cambridge)
1943 4 (26) 78
*Popular Gardening* (London)
18th June 1960 37
*Poultry Science* (Ithaca, New York)
1936 (15) 19
1963 XLII No. 4 912
Rosewarne experimental station annual report (Camborne, Cornwall)
1961 66
Rothamsted experimental station annual report (Harpenden, Hertfordshire)
1963 39
Royal Agricultural Society of England annual report (London)
1960 17–18
Royal Dublin Society, *Economic Proceedings*
1942 (3) 150
1946 (3) 273–91
*Science of Food and Agriculture, Journal of* (London)
1958 (9) 163
*Scottish Agriculture* (Edinburgh)
1949 (29) 2, 105
*Seaweed News* (Maxicrop Ltd., Holdenby, Northampton)
*Soil Science*
7th International Congress 1960 IV 72 579

*Soil Science* (Baltimore, United States)
1963 (95) 105
*Veterinary Record* (London)
1960 (72) 27, 616
1962 (74) 860, 1,041
*World Crops* (London)
1964 16(2) 40
*World's Poultry Science Journal* (Ithaca, New York)
1954 (10) 33

# INDEX

Note: Brief mentions of plants (e.g. 'Pineapple,' 'Poinsettia') imply also the effects on named plants of seaweed products, as fertilizer, mulch, spray, etc.; and recommended treatments.

# Index

# Index

# Index

# Index

# Index

# Index

# Index

# Index